Nutrition Applications
Workbook

Thomas W. Castonguay, Ph.D.
University of Maryland at College Park

Karen Israel, Ph.D., R.D.
Anne Arundel Community College

Lorrie Miller Kohler
Minneapolis Community and Technical College

Elaine M. Long
Boise State University

Judith S. Matheisz
Erie Community College

Mithia Mukutmoni
Sierra College

Sharon Rady Rolfes
Nutrition & Health Associates

Julian H. Williford, Jr.
Bowling Green State University

THOMSON
WADSWORTH

Australia • Canada • Mexico • Singapore • Spain • United Kingdom • United States

Printed in the United States of America
1 2 3 4 5 6 7   08 07 06 05 04

Printer: West Group

0-534-62383-2

For more information about our products,
contact us at:
Thomson Learning Academic Resource Center
1-800-423-0563

For permission to use material from this text or
product, submit a request online at
http://www.thomsonrights.com.
Any additional questions about permissions can be
submitted by email to **thomsonrights@thomson.com.**

Thomson Wadsworth
10 Davis Drive
Belmont, CA 94002-3098
USA

Asia
Thomson Learning
5 Shenton Way #01-01
UIC Building
Singapore 068808

Australia/New Zealand
Thomson Learning
102 Dodds Street
Southbank, Victoria 3006
Australia

Canada
Nelson
1120 Birchmount Road
Toronto, Ontario M1K 5G4
Canada

Europe/Middle East/South Africa
Thomson Learning
High Holborn House
50/51 Bedford Row
London WC1R 4LR
United Kingdom

Latin America
Thomson Learning
Seneca, 53
Colonia Polanco
11560 Mexico D.F.
Mexico

Spain/Portugal
Paraninfo
Calle/Magallanes, 25
28015 Madrid, Spain

CONTENTS

Module A: Group Diet Analysis Activities 51

By Lorrie Miller Kohler—A versatile set of diet analysis activities complete with worksheets, covering general and specific nutrition concepts, and designed for small cooperative groups. Activities 6-10 may be completed using either personal one-day Diet Analysis Plus reports or the included sample reports.

Module B: Diet Analysis Plus Exercises 123

By Elaine M. Long—Five of the eight exercises in this set require estimations and calculations regarding the nutritional properties of sample menus (for which Diet Analysis Plus reports are provided). The remaining three exercises involve an exploration of personal reports for three days' intakes plus the average of the three days.

Introduction for Instructors

Welcome to Wadsworth's *Nutrition Applications Workbook*, and thank you for choosing our nutrition texts for your course! The purpose of this supplement is provide you with a selection of activities, worksheets, and project assignments that will: (1) help you prepare for your course, and (2) engage and challenge your students by encouraging them to apply what they are learning about nutrition in practical and personal contexts. A wide range of assignments covering similar topics are provided to allow you the freedom to choose those that best fit your course. To assist you in identifying the most useful materials, the table of contents includes brief descriptions for each item or set of items, and guides to the case studies and diet analysis activities organize them topically.

This workbook is divided into two parts:

Case Studies: Twenty case studies illustrating the impact of nutrition on health, covering individual nutrients, weight management, food safety, and many other topics. A worksheet with discussion questions follows each case study.

Diet Analysis Projects and Activities: Five sets of diet analysis activities with worksheets are included. Many of these exercises are entirely self-contained and may be completed using Diet Analysis Plus reports which are printed in this workbook, while others require students to keep a record of their own intake and analyze it using either Diet Analysis Plus or hand calculations with a food composition table. Detailed instructions for recording daily intake and for two written assignments based on three-day intakes are also provided.

We hope you find this workbook a useful resource, and that your students' learning is enhanced as they work through these assignments.

Case Study Assignment Key: Major Topics for Which Case Studies Are Appropriate

Case #1	Factors influencing food choices	Case #11	Fat-soluble vitamins
Case #2	Diet planning strategies	Case #12	Major minerals
Case #3	Digestion and absorption	Case #13	Trace minerals
Case #4	Carbohydrates	Case #14	Physical fitness and nutrition
Case #5	Fats	Case #15	Nutrition for pregnancy/lactation
Case #6	Protein	Case #16	Nutrition for children and teenagers
Case #7	Metabolism	Case #17	Nutrition for adults and the elderly
Case #8	Energy balance / body composition	Case #18	Relationship of diet to health
Case #9	Weight management	Case #19	Food safety
Case #10	Water-soluble vitamins	Case #20	Food and agricultural technology

Index of Minor Topics Covered in Case Studies

Case Study #1: Cultural Differences and Nutrition

Beth is a computer analyst from New York City. She is 25 years old, 5'5" tall, and, at 160 pounds, she is overweight. Beth's fast-paced day typically starts at 7 a.m. when she gets up for work. She is always in a rush in the mornings, and barely has time to grab a quick breakfast of an instant drink or breakfast bar and a cup of coffee with cream that she consumes on the subway on her way to work.

Beth works long hours at her job and has a heavy workload that keeps her tied to her desk. Around 1 p.m. every day she takes enough time to grab lunch at the deli next door. She is in the habit of getting the same foods every day, generally a prepackaged corned beef or pastrami hoagie with the works: lettuce, tomato, onion, mayo, mustard, and American cheese accompanied by a small bag of plain potato chips. Beth usually orders a 20-oz. cola, and when she is feeling really stressed by a difficult work situation, she picks up a single serving hot apple pie to eat at her desk. Apple pie is a favorite comfort food and reminds her of happy times and special family meals.

By the time Beth gets done with work it is 8:00 p.m. and she walks to the subway station one block from her office and heads home. On the subway she decides she will have her favorite quick meal for dinner, a frozen "homestyle" fried chicken with garlic mashed potatoes and a side of creamed corn for dinner. She enjoys this meal and finds the mashed-potatoes to be very comforting after a long, tense day at work.

Beth eats most of her meals as takeout food or frozen meals out of the grocery store. She is generally too tired to cook, plus she likes the taste of convenient meals. Her freezer is always full of frozen pizzas, large-portion heat-and-serve meals, and frozen burritos. Each evening she enjoys an hour or two of television, and reads over some papers from work before fixing a bowl of Rocky Road ice cream and heading to bed.

One day her company holds a blood pressure screening for all of its employees and Beth decides to participate. Beth is shocked to find out that at age 25 she has elevated blood pressure. This news puts Beth in a reflective mood. Lately she has been noticing that she is gaining weight, and lacks some of the energy she had in college. Her father died of a heart attack two years ago at age 58, and Beth decides she should make some changes in her life before she follows in her father's footsteps to an early grave.

In another part of the world, Anna is busy with her daily routine. Anna works in her family's produce store right outside of Prague, Czech Republic. She is also 25 years old and 5'5" tall. Anna weighs 150 pounds and like Beth, Anna is overweight. Anna's day typically starts around 7 a.m. Anna's breakfast is leisurely and consists of a small Kaiser roll, a slice of ham, a small tomato, orange juice, and a cup of coffee with cream. Anna likes to start the day eating a tomato and drinking orange juice because she knows that fruits and vegetables are healthy choices, and she always has a supply of tomatoes from the family produce stand.

After breakfast, Anna heads downstairs to work. The family produce shop is in the downstairs portion of their house. While at work Anna's brother carries the large boxes of produce into the store and stocks the shelves. Anna sits in a chair at the register and rings up sales. The store closes at 1 p.m. for lunch. In the Czech Republic foods are generally prepared and eaten in the home, and lunch is the largest meal of the day. Anna and her family members enjoy a lunchtime meal of soup, usually potato soup, plus roasted meat, such as pork, boiled cabbage and dumplings. She often eats bread with her meal such as rye or pumpernickel with butter.

Author: Thomas W. Castonguay, Ph.D.

After lunch, the store then reopens until 7 p.m., when Anna returns upstairs for dinner. For dinner, Anna often eats a small sandwich of sausage on a bun. With her evening meal she often has a glass of beer. Anna is generally in bed by 10:30 p.m., and tries to get at least 8 hours of sleep so she is ready for another day's work. However, three times a week she joins her friends at a local bar where they sit and talk and Anna will drink 3 or 4 tall beers. On these occasions she smokes cigarettes since most of her friends smoke too.

Recently Anna's mother has fallen ill. The doctor thinks Anna's mother has suffered a very mild heart attack and needs to rest. Anna has a family history of heart disease; her grandmother died of a heart attack several years ago. Anna is worried about her mother, and Anna realizes that she needs to make some lifestyle changes.

Questions

1. List all of the factors that influence Beth's and Anna's food choices.

2. Which of these factors do you think was most influential for Beth and for Anna?

3. What risk factors for chronic disease do Beth and Anna exhibit?

4. Looking at the food and lifestyle choices Beth and Anna make on a typical day, how might you advise them to modify their diet and daily routine?

5. How might fried chicken dinners and Beth's convenience freezer foods contribute to diet related disease?

6. Assume Beth's breakfast consists of 31 grams of carbohydrate, 11 grams of protein, and 8 grams of fat. Anna's breakfast has 51 grams of carbohydrate, 21 grams of protein, and 14 grams of fat. Calculate the total calories in each of their meals.

7. Clearly, there are differences in food choices between Beth, an American, and Anna, a Czech. What type of research design studies the difference between groups of people?

Case Study #2: Planning a Healthy Diet

In her book, *Nickel and Dimed: On (Not) Getting By in America*, author Barbara Ehrenreich wanted to understand how a single mother might manage to live on low-wages after the welfare reform bill was passed in 1996. Ehrenreich found a minimum wage job and a place to live, and attempted to eat and survive for one month in three different cities in the United States. At one point she was desperate enough to go to a Food Pantry and received an emergency food package. Following are contents similar to those of the package she received:

- 21 cups of Corn Chex cereal
- 24 cups of Grape-Nuts cereal
- 2 cups of catsup
- ¼ cup butterscotch morsels
- 1 cup gumdrops
- 2 single-serving bags of jellybeans
- 2 sweet dark chocolate candy bars
- 1 dozen sugar cookies
- 6 hamburger buns
- Six 8-ounce cans of fruit punch (with 10% real juice)
- One loaf (16 slices/25 g each) of enriched Vienna bread
- 1 box (8 small) fruit roll-ups
- One loaf (18 slices) of enriched raisin bread
- 18 ounces of peanut butter
- 16 ounces of canned ham
- 1 package (6 servings/4 bars = 1 serving) of fig bars
- Two Ritz cracker packages (4 servings/12 crackers per package)
- One 5-ounce can Swanson chicken broth
- 2 ounces of a Kool-Aid-like drink mix (makes 8 cups)

Questions

1.   Place the foods from the given list into the appropriate food group.

Breads, cereals, grain products	Vegetables	Fruits	Meat, poultry, fish and alternates	Milk, cheese and yogurt

Author: Thomas W. Castonguay, Ph.D.

2.	Evaluate your groupings. Which food groups are over- or under-represented? Give the food pantry some suggestions to better help their patrons.

3.	Evaluate the foods in the emergency food package using as many of the six basic diet-planning principles as apply. Nutrient contents of all foods listed are found in the food composition tables in Wadsworth's nutrition texts. They can also be found on Wadsworth's Diet Analysis Plus software, or online at the USDA's web site. http://www.nal.usda.gov/fnic/cgi-bin/nut_search.pl

4.	Many of the foods provided give empty kcalories. Explain what is meant by "empty kcalories" and identify the foods this term applies to.

5.	What is a healthier dessert than the sugar cookies or fig bars? Why? (Refer to the nutrient contents.) Nutrition data can also be found on diet analysis software packaged with the text, or online at the USDA's web site. http://www.nal.usda.gov/fnic/cgi-bin/nut_search.pl

6.	Pick a food off of this list and describe the health claims the manufacturer could make? Again refer to the nutrient contents.

Case Study #3: Distressed Digestion

This is Tyler's freshman year in college and he has been earning very good grades and performing well in his part-time job. However, Tyler's eating habits have been poor and he has gained a few pounds, and his expanded waist line is making his trousers fit pretty tightly. His daily routine holds clues to his recent weight gain.

Tyler begins each morning about 9:00 a.m. when he hurriedly dresses and rushes to class without eating breakfast. He does however drink a 20-oz. cola each morning for caffeine and sugar energy. All day he sits in classes or studies in the library. About four in the afternoon Tyler takes a break for some food and he usually heads to the food court and gets a large order of chicken wings with extra spicy hot sauce. He doesn't have much time though, so he rushes through the meal, drinks another large cola, and is off to work. Tyler has a job as a waiter in a popular campus restaurant.

When the restaurant closes, Tyler heads home in his car. He usually stops to get some dinner, even though it is 11:30 p.m. He typically picks up a large pepperoni pizza that he consumes in one sitting. Sometimes he follows the pizza with handfuls of cookies that he eats while lying down on the couch and watching late night television. Lately he finds it difficult to fall asleep though. Tyler has been experiencing heartburn and acid reflux at night. His stomach makes some atrocious noises and he's been feeling bloated. Tyler has tried getting up and taking antacids, but he still suffers indigestion. Recently, a friend, who has noticed Tyler's odd stomach noises and hiccupping through class, suggested Tyler do something about his diet. At first he ignored the advice, but now his stomach pains are getting so bad he is willing to try anything.

After another painful night's sleep, Tyler wakes up early, goes to the bakery, and buys two bran muffins for breakfast, thinking this is the start of his new healthy diet. To Tyler's surprise this only makes him feel worse. He suffers horrible cramps, and later that day is forced to run to the bathroom with diarrhea. The worst part for Tyler was when he was sitting in English class and flatulence hit. Poor Tyler turned bright red and decided it was time to take the rest of the day off and go home.

When he got back to his room the phone rang. It was Tyler's grandma. He mentioned he wasn't feeling too well, and Grandma, who lived nearby, came over with chicken noodle soup made with carrots and celery and lots of broth. The two had a leisurely meal together, sitting and talking at the table for hours. Tyler ate one small bowl of soup and placed the leftovers in the refrigerator for later. Tyler then got a good night sleep, and woke up the next day feeling much better. Once again, his fast-paced lifestyle plunged him back into previous eating patterns. On the way to his first morning class Tyler was drinking a 20-oz. cola.

Questions

1. Name the symptoms Tyler was experiencing initially and describe the probable causes of his ailments.

Author: Thomas W. Castonguay, Ph.D.

2.   He tries taking antacids. Is this a good idea? Why or why not?

3.   After adding bran muffins to his diet, Tyler only felt worse. Why is this?

4.   After Tyler's grandma visited and they ate some chicken soup together, Tyler felt much better. Why might this be?

5.   Tyler experiences the unpleasant and embarrassing consequence of poor nutrition while in English class. This only adds to the already high stress in Tyler's life. What was the cause of this and how can Tyler cure this problem for the future?

6.   What advice would you give Tyler about his eating habits?

7.   What do you think the long-term consequences of Tyler's eating habits might be?

Case Study #4: Simple Sugar and Complex Carbohydrate

Clara is an African American college student. She is 5'4" and weighs 170 lbs. Clara's entire family is overweight and her father was diagnosed with diabetes last month. Her uncle has been suffering from diabetes since he was eight years old. Clara fears that if she does not change her diet she may develop diabetes as well.

On an average day Clara eats two jelly doughnuts for breakfast, chicken nuggets with sweet and sour sauce and a large soda for lunch, and goes to the drive-through to grab a super-sized burger, fries, and a large cola. She usually snacks on jellybeans during the day and is always sipping on a cola.

Clara is concerned about her diet and decides to ask her friend Mary, nutrition major, for advice. Mary gives Clara some suggestions. First she tells Clara that, while her diet is very high fat, it is also very high in carbohydrate. "I should try that diet advertised on TV. If I cut all carbohydrate I can lose 50 pounds this month!" Clara announces. Mary rolls her eyes. Mary explains that complex carbohydrates are good, but too much simple sugar is not. Mary tells Clara that she should snack on fresh fruits instead of jellybeans, and eat more starch and fiber. Plain baked potatoes (instead of fries) and fortified cereals (instead of high-fat doughnuts) would be a lot healthier. Small changes like these could improve her diet dramatically.

"Try not drinking so much soda," Mary suggests. "Bottled water or a glass of milk would be much more beneficial."

"I can't drink milk!" Clara says. "Last time I had a glass of milk my stomach hurt for hours. I think I'm lactose intolerant since it runs in my family."

Clara and Mary discuss cow's milk alternatives and Clara thanks Mary for the advice and goes home to plan her new healthy diet.

Questions

1.   Diabetes clearly runs in Clara's family. Differentiate between her uncle's and her father's diabetes. How do Type 1 and Type 2 diabetes differ?

2.   What form of diabetes is Clara at risk of? What is putting her at risk?

Author: Thomas W. Castonguay, Ph.D.

3.   Clara thinks she can't drink milk because she is lactose intolerant. How might Clara include milk and milk products in her diet?

4.   If Clara does not want to give up soda entirely, but still consume less sugar, she could drink diet sodas. What are the pros and cons of this switch?

5.   Clara discovers that a lot of her calories come from carbohydrate. She initially decided to eliminate all carbohydrate from her diet, but Mary tells her that this is a poor decision. Why?

6.   Although there may be the same amount of sugar in fruit as some candies, why is fruit a much better choice for Clara?

7.   Mary suggests eating more fiber. How would this be beneficial in Clara's case?

Case Study #5: Not Too Much and Not Too Little: Understanding Fats in Foods

Jenny and Travis are four months into their freshman year of college. The two have been friends since grade school, and find a great deal of comfort in being able to get together for dinner.

Travis loves food, but rarely tries new things. At first, the unfamiliarity of the school dining hall was scary, but once he discovered that he could get a steak and cheese sub seven days a week he was content. For Jenny the adjustment was much more difficult. Jenny has been a semi-vegetarian for many years, and the vegetarian entrees cooked at the dorm are not particularly inspired. Although she eats fish, the cafeteria rarely serves fish that isn't breaded and deep-fried. Typically Jenny sticks to salads with low-fat dressing, low-fat yogurt, or a bowl of cereal with low-fat milk for dinner.

"I'm starving!" she told Travis one night at dinner. "There is never anything I want to eat in the dorm and I'm losing a lot of weight. This winter the colder temperatures are bothering me more than usual and I have been so tired. I can't wait for this weekend. I'm going home to have a home cooked meal." Jenny couldn't help but smile as Travis's eyes lit up at the mention of a home cooked dinner. "You want to come?" she asked, knowing the answer would be a resounding, "Yes."

When they arrived the house was filled with the wonderful aromas of home cooking. There were hot rolls, salad with olive oil and vinegar, grilled salmon, and fresh chocolate chip cookies for dessert. They sat down to dinner, and after just one bite Travis decided this was the best meal he had ever had. He didn't hesitate to ask for a second helping of everything. Jenny ate a big meal as well. As she spread some margarine on her roll, she couldn't help but smile seeing her best friend enjoying his dinner.

Then came dessert. Jenny's parents served the chocolate chip cookies with some low-fat frozen yogurt, both of which were delicious. "Now, I made the cookies with real butter," Jenny's father reminded his wife, who had been cautiously watching her cholesterol. She was stuffed after a big meal anyway, so she skipped the cookies. Still, she enjoyed watching the kids devour their desserts. "I can't thank you enough," Travis remarked as they were leaving to go back to school. "Any time you want to come over for dinner, Travis, you are always welcome," Jenny's mother replied, and Travis fully intended to take her up on the offer.

Questions

1. Travis eats steak and cheese subs every day of the week. Explain the types of fats, as well as the properties of each type, that are found in these subs.

Author: Thomas W. Castonguay, Ph.D.

2. Jenny eats low-fat salads, yogurt, or cereal at every meal, and she is feeling tired and more sensitive to cold temperatures. Explain what dietary factors may have caused some of the symptoms she experienced.

3. Jenny and Travis ate salmon for dinner. What types of fat are found in this food? How does the fat found in salmon differ from that found in steak and cheese subs?

4. Olive oil and vinegar are used to flavor the salad. Explain some of the benefits associated with olive oil.

5. Jenny spread margarine on her roll, although the cookies were baked with butter. Compare and contrast these two fat sources.

6. Jenny's mother has been watching her cholesterol. She was warned not to eat the cookies because they were made with butter, which contains cholesterol. However, chocolate chip cookies contain other sources of cholesterol. What are they? (Ingredients: Flour, sugar, butter, eggs, milk, vanilla, baking powder, milk chocolate chips.)

Case Study #6: Vegetarian Diets

Dawn is a 22-year-old college senior. She is a vegetarian who believes strongly in animal rights. So, over the last three and a half years she has eliminated all meat and animal products from her diet. Dawn has also lost a little more than 15 lbs. Lately she feels that she has lost too much weight. Dawn is 5'3" and weighs 105 lbs. Her diet includes alfalfa sprouts, legumes of various types, and tofu that has not been fortified. She especially loves black bean soup, peanut butter sandwiches with multi-grain bread and several soy products. Dawn occasionally drinks soymilk. The only vegetable oils that she uses are safflower oil or canola oil. She does not take vitamin or mineral supplements, such as calcium, because she believes they are not necessary. Her theory is that she can obtain all her vitamins and minerals from the food she eats. Dawn drinks no alcoholic beverages and does not smoke. She believes it is best to eat only "natural" foods. She eats no bleached flours and eats whole grain cereals and bread. However, she makes sure to eat plenty of fresh fruits and vegetables at every meal.

Despite her health-conscious measures, Dawn catches colds easily and has a harder time fighting them off. Dawn has never been very active, but lately she has lacked energy. Dawn wonders if she should look into her diet. She decides it is time to investigate the facts about vegetarian diets in more detail, with the hope that she can feel better through more informed vegetarian eating.

Questions

1.   What kind of vegetarian is Dawn? Name and describe some other types.

2.   Examine Dawn's diet and give her some advice on how to eat a healthier diet without giving up her strong beliefs about animals.

3.   Dawn eats significant quantities of sprouts, peanut butter, and beans. What are some of the advantages and disadvantages of these protein sources?

Author: Thomas W. Castonguay, Ph.D.

4.   Name the sources of calcium in her diet.

5.   Dawn feels that vitamin/mineral supplements are unnecessary; do you agree?

6.   Aside from vitamin/mineral supplements, some people take protein and amino acid supplements. Explain some of the dangers associated with this.

7.   By eating better, Dawn hopes to gain a few pounds. Calculate the amount of protein Dawn should eat per day, if she is 5'3" tall and hoping to weigh 121 lbs.

8.   Design two one-day meal plans for a 5'3", 121-pound female who is very lightly active. Make one diet meat based and the other a vegan diet appropriate for Dawn. Compare the two in terms of calcium, iron, protein, and total calories. Use a food composition table from a textbook, Diet Analysis Plus software, or the USDA's nutrient database at http://www.nal.usda.gov/fnic/cgi-bin/nut_search.pl .

Case Study #7: Feasting and Fasting

Kathleen is a 5'6", 130-pound, 20-year-old college junior. Over the last few years she has gained then lost ten to fifteen pounds several times. Recently she has been trying hard to keep her weight down. Like many college students, Kathleen goes out every weekend and tends to overindulge. During the week she lives the life of a serious student, eating very little and getting 7-8 hours of sleep every night. By Friday she is ready to have some fun. She usually goes out with friends to a nearby burger or taco shop and then finds a party. At the party Kathleen will have four or five drinks and munch on chips or pretzels. By 3 a.m., when she and her friends are ready to call it a night, they usually are hungry again. This means finding pizza, waffles, or any other food they can find at that hour of the morning. After a late night out, Saturdays are spent catching up on some sorely needed sleep. By Saturday night she is ready to go out and do it all over again. Sundays are usually spent relaxing at a hearty Sunday brunch, watching movies with her friends accompanied by a giant bowl of popcorn, and finishing homework that's due Monday morning.

Kathleen realizes that her weekend binges may cause her to gain weight, so she cuts way down on calories Monday through Thursday. Kathleen has been trying to stay active and build up some muscle mass by running. Recently, her strict dieting is making this more difficult. During the week Kathleen eats so little that she often feels weak or lightheaded, especially following her long distance runs. Although she is running a great deal, her muscles are not getting as large as she had hoped. On Mondays and Tuesdays her diet is very hard to stick to. Kathleen is always starving. However, by the end of the week she no longer feels so hungry. Kathleen also notices that she has a much harder time paying attention, is sensitive to the cold temperatures, and finds it all too easy to catch a cold or flu.

For a while she was able to overcome this and still go out on the weekends, but she is starting to feel that it is not worth being so miserable during the week, no matter how much fun her weekends may be. Kathleen knows she is putting her body through a lot of stress, but needs someone to explain what is happening as she experiences the consequences of a weekly roller coaster ride from feasting to fasting.

Questions

1. After Sunday brunch Kathleen's digestive system is filled with nutrients. The liver receives these nutrients first. Describe what the liver does with the carbohydrates, fats, and proteins in Kathleen's Sunday brunch.

2. On the weekends Kathleen drinks substantial amounts of alcohol. Explain alcohol's effects on the liver, both short term and long term.

Author: Thomas W. Castonguay, Ph.D.

3. During periods of fasting, how are metabolic fuels used differently compared to their use with a healthy, consistent diet? Describe some of the negative effects associated with this.

4. During periods of feasting, how are metabolic fuels used differently compared to their use with a healthy, consistent diet?

5. Kathleen has a lot of warning signs that she is not eating well. Name her symptoms and explain why she feels this way.

6. Kathleen's weekend diet consists of many more calories than during the week. Discuss the body's reaction to excess carbohydrate, protein, and fat.

7. What are the long-term consequences of Kathleen's eating pattern and alcohol consumption on her vitamin status?

Case Study #8: Satiation and Appetite

Gina has been head chef at a local Italian restaurant for the last 15 years. She is from a large Italian family that loves not only cooking, but also eating. Although Gina has grown up around large amounts of food and hearty appetites, she has learned to control her weight and eat moderately. However, Gina's husband Anthony has difficulty moderating his eating habits. They have been married for 3 years and in that time Anthony has put on 40 pounds, most of which went straight to his belly. As a software analyst, Anthony gets very little exercise. To make matters worse, Anthony often has large, rich meals with his in-laws. Last week they celebrated Anthony's 55th birthday with pasta in cream sauce, breaded veal, and lots of garlic bread drizzled with olive oil. They washed the meal down with several glasses of wine, and devoured cannolis for dessert. On this birthday, Anthony realized that he really needed to change his eating habits. He is now 55, about the same age his father and grandfather both died of heart attacks, and in addition he has recently learned that both of his brothers have high cholesterol. Anthony has never had his blood cholesterol checked, but wouldn't be surprised if his levels are high too.

Anthony went to his family doctor for a physical. Anthony weighed in at 215 pounds, substantially overweight for his 5'11" frame. The doctor asked Anthony to describe his daily routine and eating habits.

Anthony reported that he sleeps late and needs to rush to work. He rarely eats breakfast although he drinks coffee with cream and sugar. Around noon everyday Anthony's coworkers go to the sub shop across the street. Anthony never joins them because he says the subs are not "real" food. Anthony does not eat due to boredom or anxiety; he eats because he loves food. This is why he brings a freshly made sandwich from home. His sandwiches are usually made of the leftover chicken parmigiana or veal cutlets from the night before. Anthony brings very small sandwiches to merely tide him over until the evening's feast.

When Anthony gets home he likes to relax by helping to prepare the evening meal. Smelling the delightful aromas and looking at the mouth-watering ingredients only increase Anthony's appetite. Dinner is the meal he lives for. When he sits at the table in the evening, he leaves the day behind and has nothing to do but enjoy a good meal and go to bed.

His typical supper consists of a tossed salad with an olive oil and vinegar dressing followed by pasta, with tomato sauce and grated cheese, or cheese sauces. Fresh mozzarella with sliced tomatoes is usually served as well. Anthony likes Italian sausage or veal with his pasta. Anthony fills his plate with pasta, has a second helping of sausage, and can't resist a bowl of gelato, which is always found in their freezer, for dessert. Anthony knows he needs to change his eating habits, but refuses to deny himself the foods he loves, so he asks the doctor for some advice.

Questions

1.	The doctor begins by determining Anthony's BMI. Calculate Anthony's BMI, recalling that he is 5'11" and 215 pounds.

Author: Thomas W. Castonguay, Ph.D.

2. What would a healthy BMI for Anthony be? What is the corresponding weight range?

3. Knowing that Anthony does not engage in much physical activity, estimate his daily energy output.

4. We know a little about Anthony's family history; why is this particularly alarming?

5. If Anthony does not change his diet soon, what health risks, in addition to cardiovascular problems, does he need to think about?

6. Anthony is easily able to overcome hunger signals during the day. What then does influence his food intake?

7. Although satiation sets in about 20 minutes into a meal, Anthony usually does not respond to this. What changes to his diet could influence the satiation he feels?

8. Anthony was much thinner just three years ago. Although much of his weight gain has to do with his new eating habits after marriage, other factors contribute to it. What are some of these factors?

9. What general advice should the doctor give Anthony regarding his diet?

Case Study #9: Diet Strategies for Overweight and Obese Individuals

Mickey and Jim are brothers. Their parents, along with most other members of their family, are overweight and have hypertension. Mickey is the youngest of the family and at age 24 is overweight. He is 5'10" tall and weighs 188 lbs. Mickey is active, although not as active as he should be. He works at an advertising agency and is on the company softball team. They have been practicing every Saturday for the last three months. From 11 .m. – 2 p.m. Mickey is running bases, hitting balls, and playing the outfield. Afterwards, he and his friends go to the local pub for a few beers and a bite to eat. This "bite" is usually a burger and fries.

Mickey's brother Jim is not active at all and is 6' tall and weighs 338 pounds. Jim is a middle school history teacher. Most of his time is spent behind a desk or sitting at the table grading papers. When he needs a break he usually goes to the fridge for a snack. Having grown up in the same household, Mickey and Jim have similar food preferences. Both love old-fashioned eating: bacon and eggs, ham and cheese, steak and potatoes. As kids, neither brother wanted to eat his vegetables, although they do eat some vegetables now. One habit they fully enjoy and never grew out of is snacking on milk and cookies. To this day they both have milk and cookies before bed. They were pudgy children, although they did not seem to eat much differently from the other kids in the neighborhood; they just gained more weight than the other kids on the block.

Mickey loves food just as much as his brother, but he knows his weight is not healthy and that his brother's is even worse. So, he asks Jim to go on a diet with him. At first Jim seemed reluctant, but now he has turned Mickey's weight loss challenge into a brotherly competition. Both are determined to lose at least 45 pounds in the next 10 weeks.

Mickey decides that he needs to exercise more to lose weight. In addition to his Saturday softball practice, he plans to jog for 15 minutes twice a week. This does not produce results quickly enough, so Mickey tries using diet pills, but only "natural" ones. Mickey tries St. John's wort but has tremendous difficulty sleeping at night. He also takes dieter's tea but can't deal with the nausea and cramping. Finally he decides to go to his doctor and get prescription drugs. The doctor tells him that sibutramine and orlistat are the most common drugs on the market, but he feels orlistat is the better choice for Mickey.

Jim also goes to his doctor, but does so before attempting to change his activity level or trying natural diet products. He asks his physician if he would be a good candidate to have his stomach reduced in size by surgical stapling. Jim schedules an appointment for a consultation with the surgeon, but never sees it through. The brothers have a long talk and decide they should change their diets and eating habits to obtain healthier, more long-term goals.

Questions

1. Losing 45 pounds in 10 weeks is not realistic. A more appropriate goal would be to lose 10% of their body weight in 6 months (24 weeks). How much weight should Jim and Mickey aim to lose?

2. As children Mickey and Jim did not eat any differently from other children but gained more weight. Why is this?

3. While growing, Mickey and Jim were overweight. How could this affect their current weights?

4. At first, both brothers set unrealistic goals, something that often leads to yo-yo dieting. Explain why this is detrimental to one's health.

5. Mickey initially tried St. John's wort and dieter's tea but experiences unpleasant effects. What are these effects and why did they occur? What about the St. John's wort could be especially dangerous?

6. Mickey is substantially overweight, to the point that his doctor was willing to prescribe medication. Why did the doctor recommend orlistat but not sibutramine?

7. Jim did not want to waste his time with conventional weight loss methods and immediately considered surgery. Calculate both Jim's and Mickey's BMIs. Based on these numbers why is Jim a more suitable candidate for this procedure than Mickey?

8. Do you think surgery is a wise choice for either of the brothers? Why or why not?

9. What behavioral changes do both Jim and Mickey need to make in order to control their weight?

Case Study #10: Absorption of Water-Soluble Vitamins

Brian is a 21-year-old college student who weighs 170 pounds and is 5 feet 10 inches tall. He lives with several roommates in a house near his campus. Brian leads a moderately active lifestyle by playing basketball and occasionally lifting weights at the campus gym. He enjoys eating pizza or burgers and drinking beers with his friends. However, Brian thinks that the pizza and burgers are not healthy foods, so occasionally he changes dietary gears. For a day or two Brian nearly fasts. However, on these fasting days he still allows himself to drink beer, and once a day he has a small bowl of cooked carrots and peas.

Generally Brian hasn't been feeling his best: he has trouble sleeping, especially when he does his "fast," and notices his gums bleed after brushing his teeth. To figure out what is causing these symptoms he looks at his food intake on a particular, and typical, Saturday. It is as follows:

Breakfast: 10:00 a.m. (at his house with friends)
Two frosted strawberry Pop Tarts
Super 20 ounce Gatorade

Lunch: 1:30 p.m. (at a nearby McDonald's)
Big Mac
Supreme Super Size fries
Super Size Coke
Apple pie

Snack: 5:00 p.m. (at his house with friends)
Two cans Bud Light beer

Dinner: 7:00 p.m. (at home)
Three slices of Pizza Hut Supreme pan pizza
Eight cans Bud Light beer

Snack: 1:30 a.m. (at home)
One slice Pizza Hut Supreme pan pizza

Here is an analysis of his diet:

Name:	Brian					
Analysis of:	1 day (plus comparison to RDA)					
Nutrient	Amount	% RDA	Nutrient	Amount	% RDA	
Calories	4446	153	Vitamin C	51.5 mg	86	
Protein	114 g	197	Vitamin D	1.06 mcg	11	
Dietary fiber	18.9 g	---	Vitamin E	21 mg	210	
Vitamin A total	424 RE	42	Calcium	1600 mg	133	
Thiamin—B$_1$	3.18 mg	212	Iron	24.8 mg	248	
Riboflavin—B$_2$	3.29 mg	194	Magnesium	361 mg	103	
Niacin—B$_3$	45 mg	237	Phosphorus	1670 mg	139	
Vitamin B$_6$	2.41 mg	121	Zinc	12.9 mg	86	
Vitamin B$_{12}$	5.2 mcg	260	Alcohol	113 g	---	
Folacin	235 mcg	118				

Exchanges			
Bread	20.8	Vegetable	2
Meat	6	Milk	2
Fruit	1.6	Fat	36

Calorie Breakdown	% kcal
Protein	10
Carbohydrate	45
Fat	27
Alcohol	18

Author: Thomas W. Castonguay, Ph.D.

Questions

1.   Which water-soluble vitamins does Brian have too much or too little of?

2.   Would you recommend that Brian take supplements for the vitamins he lacks in his diet? Why or why not?

3.   Why does Brian have trouble sleeping, especially when he restricts his diet?

4.   What can he do to stop his gums from bleeding so easily?

5.   How might Brian's alcohol intake be affecting his nutrition status?

Case Study #11: Environments and Susceptibility to Vitamin D Deficiency

Sharon and Janette are sisters. While in high school they each decided to adopt vegan diets. Sharon and Janette plan their diets carefully and do not take a nutritional supplement, preferring to get their nutrients from food. Neither drink fortified juices or soymilks, again, preferring to get nutrients "naturally."

Sharon recently turned 30. She has two young children and is pregnant with her third child. Her husband, Russell, is in the military, and a few years ago the family moved to a base in Alaska. Sharon does not like the cold weather and long, dark winters. She has been working in a local factory and does a lot of physical labor. Recently, Sharon has noticed a lot of pain in her lower back and legs. It is far too early to be a result of her pregnancy, and Sharon is becoming concerned.

Janette, who moved to southern California after college, is the younger of the two sisters. She was married in California, started a family, and is embarking on a new career. Janette works the night shift on a newspaper as a copier. Her job is to assure that the paper comes off the press and is ready to be delivered first thing in the morning. Janette's husband is able to take care of their son while Janette sleeps during the daytime. They are discussing having another child, but Janette does not think that her inverted day-night schedule will be suitable for her growing family. She also fears for her own health if she has another child. Janette has had several bone fractures in the last few years, and recently has had a lot of pain in her pelvis. Although she wants to have another child, she needs to be examined by her doctor first.

Sharon and her sister talk regularly on the telephone. They look forward to comparing lives as well as ailments. Both of them know that something needs to be done to better understand their health problems.

Questions

1.   What might be the cause of Sharon's leg and lower back pain?

2.   What might be the cause of Janette's pelvic pain and easily fractured bones?

Author: Thomas W. Castonguay, Ph.D.

3. Is there anything Sharon could do to help her ailments?

4. Is there anything Janette could do to help her ailments?

5. Each sister is having or is considering having another child. Why could pregnancy and repeated periods of lactation have adverse consequences on their health?

6. Are the children of these sisters at risk of similar health problems? If so, name the most probable diet-related disease they will develop.

7. Which child is at greater risk and why? What should these mothers do to prevent their children from developing dietary health problems?

Case Study #12: Hidden Sodium

John is a 35-year-old advertising executive. He is fairly active, has a BMI of 23, and is in generally good health. However, at his yearly physical his doctor discovers that John has high blood pressure. John assures his doctor that he eats well, and can't understand why the numbers are so high. John's doctor tells him that he probably is salt sensitive, and needs to cut back on the sodium in his diet. When John returns home he immediately puts the saltshaker in the back of the cupboard, so he is not tempted to add salt to his foods. He also replaces his usual snack of salted peanuts with dry cereal, his favorite being cornflakes. He has heard about packaged foods being high in sodium, so he swears off his usual frozen dinners and takes the time to prepare some of his favorite meals. John eats Mexican rice and beans, or a piece of smoked halibut with mashed potatoes to go with it. Since lunch is usually at the office, John decides he needs to start bringing lunch instead of ordering Chinese food with his coworkers. John goes out to the store and buys a package of bologna and some American cheese to make sandwiches. He enjoys a pudding cup for dessert. He is always very thirsty so he brings both a bottle of water and orange juice to sip on, but rarely milk. After a month on this new diet John returns to his doctor, assured that his blood pressure has gone down. To his surprise, the numbers have not budged. "I don't understand! I don't put salt on anything," John tells his doctor.

Questions

1. Explain how sodium is the contributing factor to John's hypertension.

2. The first step John takes in lowering his sodium intake is hiding the saltshaker. Do you think this was very effective? Why or why not?

3. John replaces his snack of salted peanuts with corn flakes. Using a food composition table, look up the sodium content in both and explain why this was or was not a good choice.

Author: Thomas W. Castonguay, Ph.D.

4. John is aware enough to know that packaged foods and Chinese food can both be extremely high in sodium. What makes these foods high in sodium even if they are not particularly salty in taste?

5. He replaces frozen dinners with Mexican rice and beans or smoked halibut, two seemingly healthy choices. Compare their sodium contents to the RDI.

6. The sodium in John's diet makes him thirsty. Explain why this is.

7. John generally drinks water or juice to quench his thirst, but rarely milk. Although he does have some dairy from pudding or cheese, he does not ingest much. How could this, in conjunction with his high sodium levels, have adverse effects on his health?

8. Design a diet for John that would lower his sodium intake, improving his blood pressure.

Case Study #13: A Little Is Good, but More May Not Be Better—Iron Deficiency and Toxicity

Laura is a twenty-one-year-old communications major at a small college just outside of Duluth, Minnesota. She is an average student in the classroom, but excels in intramural track and field competitions. She is about to start her senior year, and Laura wants to maintain the school track and field records, which she has earned in the past few years. Because she wants to be in top physical condition when the school year begins, Laura changes her diet so that she now only eats low-fat, highly nutritious foods. She has become a vegetarian. Although she still eats eggs and dairy products, she now chooses soy patties over hamburgers. Laura also begins running every morning before school.

She was in the best shape of her life the first few months of the school year, winning all of her events. Recently, however, Laura has not been feeling too well. She has noticed that it is much harder to gather the energy for class every day. She has a hard time "getting up" to start new things. At first, she noticed it only once a month, while she was menstruating, but now she feels tired and weak more often. She gets particularly tired after running sprints. At the beginning of the year she set the pace for her team, but now she finds it difficult to keep up.

At first, she thought she was just tired and stressed from such an active lifestyle, but now she is noticing other symptoms, causing her to be a bit more worried. She is now experiencing shin splints (persistent pain in the shins). Lately she has looked pale. She gets frequent headaches and finds herself much more irritable. This may be because she has been having trouble sleeping lately. This winter has also been particularly cold and Laura can't seem to warm up. Laura's symptoms were not severe enough for her to worry, so she put off seeing the doctor. However, Laura now tends to get more bruises than she used to. And, when she gets cut it takes much longer to heal. When she does get a cut it often becomes infected. Laura decides it is finally time to see her physician.

While at the doctor she has blood drawn and analyzed. Her results are sent in the mail later that week. She receives the following information:

Laura's Test Results

Hemoglobin	110 g/L
Hematocrit	.30
Mean Corpuscular Volume	70 FL
Mean Corpuscular Hemoglobin	25 pg
Serrum Ferritin	10 ug/L
Erythrocyte Protoporphyrin	1.36 umol/L RBC

Lower Limits of the 95% Reference Range for a 20—44-Year-Old Female

Hemoglobin	120 g/L
Hematocrit	0.35
Mean Corpuscular Volume	80 fL
Mean Corpuscular Hemoglobin	27 pg

Recommended Cutoff Value for Confirmatory Test for Iron Deficiency for Someone 15 Years or Older

Serrum Ferritin	Less than 12 ug/L
Erythrocyte Protoporohyrin	Greater than 1.24 umol/L RBC

*P.R. Dallman. 1987 Iron deficiency and related nutritional anemias. In: Hematology of Infancy and Childhood 3rd ed. (D.G. Nathan and F.A. Oski eds.) pp. 274-296, Saunders, Philadelphia.

Laura can tell that her iron is low, so before discussing it with her doctor she immediately goes to the store and buys iron supplements. She takes a 60 mg supplement before every meal and again before going to bed. Laura makes sure to swallow the supplement with a glass of orange juice because she knows Vitamin C enhances iron absorption. She also tries to eat more iron-rich foods, without changing her vegetarian diet. She loves spinach, and, having read that it is high in iron, finds a way to incorporate it in most every meal.

Laura quickly begins to feel better, but this doesn't last for long. After one month her symptoms return and additional symptoms develop. Laura has noticed that she was constipated when she first began increasing her iron, but lately she has had nausea and diarrhea. She feels just as weak and lethargic as before, and her occasional cuts and scrapes continue to be easily infected. Laura begins to wonder: if it's not an iron deficiency, then what is wrong with her? She knows something has to change, but needs help figuring out what to do.

Questions

1. Why might Laura's symptoms have returned, even after changing her iron consumption?

2. Laura experiences many symptoms that eventually bring her to the doctor. Which of these are indicative of iron deficiency?

3. At first Laura only experiences symptoms during menstruation. Explain why this is.

4. Interpret the results of Laura's bloodwork.

5. When Laura discovers that her iron is low she immediately makes dietary changes. What misconceptions does she have about iron intake and absorption?

6. What general advice can you give Laura to improve her health?

Case Study #14: Can Physical Fitness Come in a Bottle?

Ryan is a 21-year-old college student who is very interested in bodybuilding. His ambition is to compete in bodybuilding on the local and then state levels, and perhaps one day enter competition on the national level. He is 6'1" and weighs 212 lbs. Because of his lean body mass, he does not look overweight. He works out every day for two to three hours, working different body parts each time. His goal is to gain 25 pounds. To do so, he is taking dietary supplements, including protein, vitamin E, and creatine. The gym he attends has high-quality equipment and support services. Among the gym's services are cardiovascular and respiratory evaluation, fat analysis, and dietary instruction. He considers speaking with the dietitian, but her main job is working with people who are overweight, so Ryan decides not to seek her advice.

What Ryan knows about nutrition comes from various bodybuilding magazines. He has put himself on a high-protein, low-carbohydrate, low-fat diet. He obtained a calorie counter book from a newsstand and uses it to determine the amounts of carbohydrate, fat and protein he eats. He has discovered that he takes in 10 percent or less of his total calories from fat. Ryan is eating 150 g of protein per day. He understands that the best form of protein is egg, so his 150 g of protein includes four raw eggs per day. In addition, he eats 3 g of protein powder before working out, 3 g after working out, 3 g in the morning, and 3 g in the evening. The protein powder is a combination of amino acids he adds to skim milk. He is also taking a high-stress multiple vitamin and mineral tablet every day. The tablet contains 200 percent of his requirement for vitamins A and D. The vitamin A is in the acetate form. In some of his magazines he has read that vitamin E prevents oxidative stress caused by physical activity, so he takes a Vitamin E supplement each day. He has also read that Vitamin C is necessary for protein synthesis, so he takes 2 g of Vitamin C a day. Both are in addition to the multiple vitamin he is taking.

One day, Ryan decided to use the gym's service of bioelectrical impedance testing to measure his body fat. To his surprise Ryan discovered that he has 10 percent body fat. He was disappointed with this and wanted to get his body fat down to 7 percent. To do so he decides to begin a cardiovascular fitness program. He begins jogging each morning, but quickly feels light-headed and tired. He doesn't understand why, because he is able to do weight training exercises for hours without any adverse effects. Ryan decides to ask one of the trainers at the gym about this. The trainer suggests it would be a good idea to talk with the gym's dietitian.

Questions

1.   What is your opinion of Ryan's current diet? Why?

2.   Ryan is on a high-protein, low-carbohydrate, low-fat diet. Does this help or hurt his goal of gaining 25 pounds? Explain.

Author: Thomas W. Castonguay, Ph.D.

3.   Ryan is taking a multi-vitamin as well as extra supplements and a variety of foods. Refer to the information on water- and fat-soluble vitamins in your textbook. What adverse effects can these extra supplements have?

4.   Ryan is lifting weights for several hours a day. What type of exercise is this? Explain what this has to do with a fuel source and protein use.

5.   Why does Ryan's diet give him energy for weight lifting but not for jogging?

6.   Do you think Ryan's goals of gaining 25 pounds and reducing his body fat to 7% are realistic and healthy? Why?

7.   Describe a better diet and exercise program for Ryan.

Case Study #15: Nutrient Needs of Women and Infants

Cameron and Bradley are 25-year-old, budding Hollywood stars. They live in a large house where they host parties almost every week. Recently, Cameron decided that she is ready to become a mother. She and Brad want only the best for their child and are willing to change their lifestyle to help ensure the arrival of a healthy baby. Cameron goes to her doctor to get some advice. First, the doctor weighs her and finds that she is 5'4" and only 100 lbs. He asks about her diet and finds that she lives mostly off of coffee and cigarettes during the day and wine and liquor every night. Most of her caloric intake from food is eaten as hors d'oeuvres. She is physically inactive, never getting any exercise. The doctor can't help but be critical, and tells Cameron that if she is serious about ensuring the health of her baby she will need to change her lifestyle drastically before even attempting to become pregnant.

When Cameron returns home from the doctor she and Bradley discuss the matter. Having a family is very important to them, so they agree make changes right away. First, they both quit smoking. They also stop throwing parties and drinking alcohol. Cameron focuses on improving her diet. She eats three balanced meals per day, stops drinking coffee, and manages to put on a few pounds, bringing her to a healthy body weight (an additional 10-15 pounds). She is particularly deliberate in making sure that she eats several servings of fruits and vegetables every day.

A few months later she goes back to her doctor. He is impressed with her dedication and tells her that she is now ready to conceive a child who is much more likely to be born healthy. Only one month later a pregnancy test turns up positive. Her doctor confirms that she is pregnant and prescribes prenatal vitamins. He strongly urges her to continue her healthy habits.

Over the next few months Cameron goes through the typical experiences of being pregnant: nausea, constipation. But, on the positive side, she has shown a healthy weight gain. Food cravings emerge, and Brad is regularly sent to the store for pickles, ice cream, apple juice, and potato chips.

As her nine months come to an end, Cameron begins to consider whether or not she will breastfeed her baby. She is not sure if this is a good idea, so she calls her sister Annie in Utah. Annie is very different from Cameron. She lives a relatively simple life, and has two healthy children of her own. Cameron asks Annie if she breastfed her children. Annie tells her that she did not breastfeed, but if Cameron wants to there are certain things she must do first. She tells Cameron that she must take vitamin and mineral supplements, particularly vitamins D, K, and calcium. Cameron tells her sister that the doctor has put her on prenatal vitamins, but Annie insists that her requirements will change once she is breastfeeding, so Cameron will have to adjust her diet.

Questions

1.   If Cameron had not changed her habits before conceiving a child, what are some of the detrimental effects that could have occurred?

Author: Thomas W. Castonguay, Ph.D.

2. The doctor prescribed a prenatal vitamin. What does this include and why is it important?

3. What are some of the likely causes of Cameron's nausea, constipation, and strange food cravings?

4. She gains a healthy but substantial amount of weight. About how much would this be and why is it so important?

5. Do you think Cameron should breastfeed her child? Why or why not?

6. Annie has several misconceptions about the needs of a lactating woman. What are they? Give Cameron appropriate advice.

7. Design a healthy meal plan for Cameron leading up to conception, during pregnancy, and after the birth.

Case Study #16: Food Choices Differ Among Age Groups

Jenna Swanson and her husband Alex live in Oberlin, Ohio with their three children, fourteen-year-old son Kevin, four-year-old daughter Alissa, and ten-month-old daughter Amanda.

Alex is a college professor and professional musician and works very long hours. He is never home to eat dinner with the kids. Jenna is overwhelmed with caring for three children on her own and thinks that it might save her some time and effort if at each meal she served the same foods to all three children.

On Jenna's first effort at serving one meal, she plans a menu of hotdogs, grapes, and apple juice. Jenna sits all three children in front of the television, a place where they spend a lot of time. She instructs Kevin, the 14-year-old, to keep an eye on his younger sisters while she cooks.

Kevin, who is starving after this afternoon's track practice, sneaks a package of peanut butter cookies, and has a before-dinner snack. Four-year-old Alissa is all too eager to help him eat the cookies. A few minutes later Jenna calls them to the table for dinner. She has prepared hotdogs with toasted white rolls and barbeque sauce, grapes, and a glass of apple juice for each.

They sit down at the table, but Alissa doesn't eat a thing. She has been eating cookies and is simply not hungry. Kevin, on the other hand, cleans his plate in minutes. He takes about two sips of the apple juice, and then goes to the fridge for some soda.

Baby Amanda is still learning to hold her cup and spoon and Jenna helps her eat. Her hotdog is cut into small pieces but she does not like the hot dog pieces with barbeque sauce on them or the bread. Amanda does, however, enjoy the grapes and apple juice. Jenna gives her a bottle of juice after dinner, as she puts her down to bed.

Now that baby Amanda is asleep, Jenna readies Alissa for bed. In doing so, she notices that Alissa is scratching all over. Hives have developed on her back and she is complaining of a headache. In a panic, Jenna calls the doctor to find out what is wrong. The doctor calls back and calmly tells her that Alissa is probably having an allergic reaction, but it doesn't sound too severe. He tells her to use cortisone on the hives and keep an eye on her to make sure the symptoms don't get worse. Later that week he wants to conduct allergy tests to find the cause of her reaction.

Alissa's reaction subsides and she is eventually able to go to sleep. By this time Alex is home from work. He and Jenna fix themselves a late dinner of pasta, with meat sauce, dinner rolls, and salad. Kevin, who is still hungry, joins them. He eats a large bowl of pasta without the sauce and several rolls, but skips the salad. Kevin then grabs another soda from the fridge, and watches some television before going to bed.

Questions

1. Evaluate the dinner that Jenna prepared for the children. What are the pros and cons of these foods for all three of her children?

Author: Thomas W. Castonguay, Ph.D.

2.	Why is Kevin always hungry?

3.	Was this an appropriate dinner for baby Amanda? Why or why not?

4.	What most likely gave Alissa an allergic reaction?

5.	What important foods does Kevin lack in his diet?

6.	What is the problem with giving Amanda a bottle at bedtime?

7.	Alex is becoming concerned that Alissa is gaining weight. Although she is only four and he does not want to restrict her diet, he fears obesity and heart disease later in her life. Is this a valid concern? Why or why not?

8.	What would more appropriate food choices be for all three children?

Case Study #17: Drug and Nutrient Interactions in the Elderly

Grandma is an amazing woman. Up until recently she was cooking, cleaning, exercising, and staying very active in her community. Unfortunately, at age 85 Grandma broke her hip. Her life suddenly became very different. She had become accustomed to her independence, but now is no longer able to get through the day without assistance. As a result, her family persuaded her to move from her home in Florida to Seattle, Washington, where she could be close to her closest relatives. Grandma was not thrilled about the move but knew it would be the best thing for her.

Once Grandma arrived, her children observed some unsettling behaviors. They saw that Grandma ate very little. She had always been physically active but says now that she is immobile, she is simply not as hungry. Although this makes sense, her children worry that she is not eating meat or dairy products, and may become ill from this. She has been experiencing nausea, dizziness, and dry mouth at times, and urinating can be very painful. As a result she never takes more than two sips of her juice at breakfast or her water at lunch and dinner. The children decide she should see a doctor. She agrees and they take her to a local physician.

The doctor advises her to make an effort to return to her former good eating habits. He also found that she had a urinary tract infection, and was about to prescribe tetracycline when he discovered that she is taking Coumadin for heart disease and aspirin for her hip. He finds an alternative medication. The family hopes that with sound medical advice and time Grandma will return to her former healthy habits and be the same cheerful woman she always was.

Questions

1.   Grandma is on many different medications. Describe some of the consequences drug interactions can have.

2.   Why does the doctor quickly change his prescription of tetracycline?

Author: Thomas W. Castonguay, Ph.D.

3.   What is the most likely cause of Grandma's urinary tract infection?

4.   What possible interaction should Grandma be aware of regarding the Coumadin she is taking?

5.   Although she claims she is simply not hungry, what could Grandma's lack of appetite be an indication of?

6.   Grandma does not drink milk, but needs calcium in her diet. How much calcium does she need, and what can she do to add this to her diet without drinking milk?

7.   Aside from calcium, milk is also a substantial source of vitamin D. Why is adequate milk consumption more of a concern than before, given Grandma's current health and recent change in living environment?

Case Study #18: Genetic Predisposition to Chronic Disease

Clyde, a 30-year-old marketing executive, has been reading a great deal lately about the Human Genome Project. Modern science is offering new opportunities to anyone who wishes to know about the genetic predispositions they may have to certain diseases. Clyde has learned that modern genetic screenings validate many people's predictions based on their family histories. Motivated by his recent reading, Clyde researches his family history and develops a list of possible health concerns.

Unfortunately for Clyde, the list of possible health complications is rather lengthy. All of Clyde's family members are overweight, including Clyde. Many of his relatives have high blood pressure. Hypertension led to his grandfather's heart attack, and two of Clyde's uncles have Type 2 diabetes. Clyde is most concerned about his genetic predisposition to colorectal cancer. Several cousins were diagnosed with this cancer while in their thirties. Clyde knows that thousands of people each year die from cancer, and decides that it is not too late to change his dietary habits in order to minimize his apparent genetic susceptibilities.

Clyde is now devoted to eating right and exercising. He restricts his caloric intake and exercises for 30 minutes 5 days per week. Clyde had always been slightly overweight, but now is very fit and able to maintain his ideal body weight. He also gets regular cholesterol screenings and limits the amount of meat he eats to help lower his cholesterol and saturated fat intake. Clyde used to drink wine or beer most nights at dinner, but now he no longer drinks more than three alcoholic drinks per week. He has never smoked, but now avoids even being around those who do. Clyde drastically limits the amount of salt he eats, while adding a multitude of fresh fruits and vegetables to his previously high protein, meat-based diet. He now eats fish regularly, and incorporates garlic or ginseng into most of his meals. Finally, he no longer spends his summer weekends working on his tan.

Questions

1.   Which dietary practices may help Clyde prevent cardiovascular disease?

2.   Which of Clyde's practices may help prevent hypertension?

Author: Thomas W. Castonguay, Ph.D.

3. How can diet and lifestyle help prevent cancer?

4. Which dietary practices may help prevent diabetes?

5. What herbal medicines does Clyde incorporate in his diet and what are their possible effects?

Case Study #19: Food Safety on the Go

Jessie lives in Wyoming, right outside of Yellowstone National Park, and spends much of her free time camping and hiking in the area. She is excited when her sister, Abbey, and Abbey's two children come to visit. Abbey's son Jordan is only a few months old, but her daughter Nicole just turned five. She is old enough to enjoy the outdoors, so Jessie invites Nicole to go hiking for the day while Abbey stays home with Jordan.

"What would you like to pack for lunch?" Jessie asked Nicole as they prepared for their hiking trip. "Cookies!" Nicole replied enthusiastically. Jessie packed cookies for dessert, tuna salad sandwiches prepared with mayonnaise, lettuce and tomato, and apples. She also packed plenty of bottled water. Jessie is always very hesitant to eat anything with pesticides or artificial coloring, so the lettuce and tomato on her sandwiches were grown in her own garden. The apples were organically grown, and they have the USDA sticker to prove it. Once she has placed them in an insulated backpack with an icepack, the two are on their way to experience the great outdoors.

They hike for hours, stopping by a stream to eat their lunches. "I'm thirsty," remarked Nicole as she headed to the stream to take a drink. Jessie stopped her and handed her some bottled water. After lunch they continued following the trail. Jessie kept a close eye on Nicole, who enjoyed picking berries she found on shrubs along the sides of the trail. Jessie thought they might be blackberries, but wasn't positive, so she did not allow the child to eat them. The two made it back home just in time to watch a beautiful sunset.

"Dinner is ready and waiting, " said Abbey as they walked in the door. Nicole and Jessie were glad to hear this because they were starving. Abbey had made baked chicken, steamed broccoli, and roasted potato wedges served with a honey dipping sauce made from raw honey and mustard. Everything looked delicious, but Jessie noticed that some of the potato wedges had a layer of green under the skin. Luckily Abbey had made more than enough potatoes, and Jessie knew enough to discard the wedges with green layers. Even Jordan seemed to enjoy the meal, although he was not allowed the honey dipping sauce. Nicole ate as much as she could and then excused herself to go to bed. She was exhausted after a full day of hiking with her aunt Jessie.

Questions

1.   Jessie only eats organically grown fruits and vegetables. Is this necessary? What are some of the pros and cons of doing so?

2.   Why was it so important that Jessie bring a cooler and ice pack?

Author: Thomas W. Castonguay, Ph.D.

3.   Jessie promptly stops Nicole from drinking water out of the stream. Why would this water not have been safe? Luckily, Jessie brought bottled water, but if she had not what could she do to prepare the stream water so it would be safe to drink?

4.   Was Jessie right to keep Nicole from eating the berries? Why or why not?

5.   Abbey prepared potatoes that had green spots. What is this an indication of? Explain why Jessie felt it was important to make a new batch, even though the potatoes were cooked.

6.   Jordan was not allowed to eat the honey dipping sauce. Why is this?

Case Study #20: Environmental Dilemma

Brette is an Environmental Studies major at her college in California. She is taking a class on the agriculture industry, and is very interested in understanding differences in farming techniques. As a class trip, the professor takes them on a tour of several farms in the area.

The first stop is in Apple Valley, right outside of Sacramento. The farmer, Glen, is an older man, who explains that he is from several generations of farmers. They have learned from one another and prefer to use traditional farming techniques. Glen has incorporated some newer technology, however. Water is often a scarce resource in California. Glen uses modern irrigation technology to conserve water in ways that were not available to his grandfather. He can focus water use during dry spells on specific areas of greatest need. His farm takes a lot of work, but he produces apples, all grown without the use of pesticides. Glen explains that the seeds he uses were genetically engineered. These "special" seeds have produced trees that are pest resistant and disease resistant, so pesticides aren't needed. The apples have a vibrant color and are better able to withstand storage and shipping. He is very proud of his ability to combine the traditional styles of his father and grandfather with environmentally safe practices available through modern technology.

Next is a visit to an almond orchard in Arbuckle, CA. The farmer, Maggie, prides herself on the sheer quantities she ships out every week. She explains that her orchard employs many workers that all get paid fair wages because the farm is so profitable. To protect the crops and ensure mass production, Maggie uses any means available to reduce pest populations. She employs mechanical trapping devices, pheromones, biological pesticides, and, if needed, chemical pesticides. Although this may involve some environmental risks, ensuring crop growth enables them to maximize the production of almonds.

The class now makes its way to Sacramento Valley. They visit a rice farm. Although large, this farm seems to have very few employees. The owner, a man named Bill, explains that he needs fewer employees than other rice farms this size, because modern technology contributes extensively to his farming practices. For example, he uses laser technology to precision level and grade fields, tractors to prepare seedbeds, and self-propelled combines for handling muddy soil. He uses global positioning satellites to pinpoint the specific parts of rice paddies that need attention. Although this technology is expensive, the farm saves money from lower labor costs. He can use less water, save fuel, and reduce the use of pesticides. Bill is able to adjust his use of resources to ensure the least waste and environmental harm, while keeping production at a maximum.

Finally, Brette and her classmates reach the Central Valley. They see acres and acres of cornfields. The farmer here is named Joseph. He explains that his corn can resist most anything. Not only is the corn designed to grow in drought conditions, but also it is "Round up" ready. "Round up" is a herbicide that he uses two or three times per year to kill the weeds that may threaten the crop. The corn itself is genetically resistant to this herbicide so although weeds are killed, the corn survives.

Once back at school the students must gather the information they have collected. They are required to classify each farm as either high-input or low-input, and compare the different farming practices used in terms of economic and environmental impact.

Author: Thomas W. Castonguay, Ph.D.

Questions

1.   In which classification would you place the first farm in Apple Valley? Why?

2.   In which classification would you place the farm in Arbuckle? Why?

3.   In which classification would you place the farm in Sacramento Valley? Why?

4.   In which classification would you place the farm in Central Valley? Why?

5.   Describe farming techniques common to several of these farms.

6.   Describe the costs and benefits of each farming technique.

7.   Among the farms discussed, which one do you think is the most environmentally friendly? Why?

HOW TO RECORD YOUR 24-HOUR DIETARY HISTORY

1.   Record **every item** that you consume in one 24-hour period.

2.   Be sure to record:

- The **amount consumed** (refer to your text's food composition tables to determine the amount of each specific food considered to be one serving, sometimes listed as "measure").
- If you ate more than one serving, record as 1.5 or 2 or 3.75, etc.
- If you ate less than one serving, record as 0.75 or 0.5, etc.

3.   Record **how the food was prepared or cooked** (fried, broiled, poached, raw, etc.).

Format:

Name of Food	Amount Consumed	How Prepared
honey-nut cereal	2 cups = 2 servings	no preparation
fortified soy milk	1 cup = 1 serving	no preparation
whole-wheat bread	1 slice = 1 serving	toasted

4.   Keep this list for 24 hours. Then enter this information into the Diet Analysis Plus computer program.

5.   When you have entered all of your foods into the computer database, print your spreadsheet and complete set of reports for that day.

6.   Repeat above steps for the number of days you have been assigned, and print the "analyze (average)" reports to see an analysis of your average intake over several days.

Author: Julian H. Williford, Jr.

THREE-DAY DIET ANALYSIS PROJECT

1. Format for report: typed, double-spaced.

2. Your report must contain the following **in order**:

   a. **Cover page** with title, name, date
   b. **Introduction:** One paragraph explaining why you are doing the analysis.
   c. **Results:** Printouts from Diet Analysis Plus: **(1)** the bar graph describing food consumption and the percentage of nutrient requirement being met for the average of 3 days, **(2)** the Food Guide Pyramid for the average of the 3 days and **(3)** the profile of the individual's requirements. (Do **NOT** include **individual** days' information.)
   d. **Discussion:** Answer the following questions in complete sentences.

      1. What was your **average caloric intake** for the 3 days? After 6 months of eating this way, how much weight will you gain or lose?
      2. How many **grams** and **calories** of carbohydrates did you consume? What **percentage** of your total calories came from carbohydrates?
      3. How much **total fat** did you consume in **grams** and **calories**? What **percentage** of your calories came from total fat?
      4. How much protein in **grams** and **calories** did you consume? What **percentage** of total calories came from protein?
      5. For those vitamins and minerals that you consumed in deficit of the RDA, **first list them** and **name one food that you would eat and the amount needed per day to meet the RDA for EACH of the vitamins or minerals in which you are deficient.** If you have **no** deficits, please address any of the following areas in which you are in **excess**: vitamin A, D, E, K, B₆, or iron.

Please **keep** all your information from Diet Analysis Plus, including each day's information as well as the averages. You may be asked to refer to it again.

THREE-DAY DIETARY ANALYSIS — WRITTEN REPORT

Record your food intake for 3 days, and enter this data into Diet Analysis Plus. With the information that you obtain from the DA+ spreadsheets and reports, answer the questions below. They should be typed on a **separate** sheet, and contain *sufficient, accurate* detail in order to receive full credit.

All questions should be answered based on your **3-day average** intake, not on your daily intakes. You should be focusing on the sheet that shows your intake for all nutrients, and your "% of Goal" for each nutrient. Before you answer the questions, make sure you have double-checked your printout, and that all of your numbers look reasonable (i.e., your kcal intake is not 7800, or your vitamin D intake is not 560% of your Goal). If you would like, allow your instructor to look over your printout before writing your report.

1. Compare your intake of **macronutrients and dietary fiber** (*only*) to the intakes recommended by the Dietary Guidelines for Americans. Do this by creating a table showing the Guideline's **percent kcal recommendations** (or grams, for dietary fiber) in one column, your percent kcals from each energy nutrient (including saturated fat) in another, and a third column indicating whether your intake meets, exceeds, or is below the recommended levels.

   <u>Note:</u> For the macronutrients, the "%" listed on the printout is the percent of your GOAL, **NOT** the percent of your kcals. You must **calculate** what percent of your total kcal intake comes from each macronutrient. Hint: you will need to convert the amounts given (grams) to kcals, and then figure out the *percent* (part/total x 100).

2. List the nutrients that you are consuming in quantities that are higher than your goal intake (over 150%) or lower than your goal intake (under 70%).

   First look at your "higher than goal" list. Which (if any) of these dietary components have known adverse health implications (i.e., contribute to nutrition-related diseases)?

   Next, look at your "lower than goal" list. Of these, which (if any) pose a health problem when consumed in quantities lower than recommended?

   <u>Note:</u> Mention only those health problems that relate to you (or the person whose intake you're analyzing). No need to say that insufficient "nutrient X" will impair growth in children if you are an adult, or that excess may cause sterility in males if you are a female.

3. Look at your "personalized" Food Guide Pyramid for your 3-day average intake. Compared to the recommended number of servings within each category, are you consuming too many, too few, or the recommended number?

   Addressing the categories from which you are getting *too many* or *too few* servings, show the changes you would make to meet the recommended servings (i.e., state what *specific foods* you would add or remove from to your daily intake). Indicate how the *specific changes* you make will bring you closer to meeting the recommended nutrient amounts in question #2 above (for example: adding 2 servings of fat-free yogurt will increase my calcium, which was too low, without raising my saturated fat and cholesterol, which was too high). **See following note regarding the format for this answer.**

Author: Karen Israel, Ph.D., R.D.

4. Compare your recommended calorie intake to your actual 3-day average calorie intake (show both). Is it about the same, higher, or lower? Based on this observation, theoretically, should you be gaining, losing, or maintaining weight? Explain why. Comment on whether or not you think a change in calorie intake is needed.

5. Do you think this 3-day dietary record can be used to make accurate assumptions about your overall, long-term nutritional status? Why or why not (i.e., what are possible sources of error, or factors that this program may not take into account)?

Some things to keep in mind when writing your report:

- For question #3, be as thorough as possible. Refer to **specific foods** on your 3-day lists, and indicate which may be contributing too many nutrients (sodium, fat, sat. fat, cholesterol, etc.) or calories, and how those foods could be modified or replaced with high nutrient/ lower calorie foods. For example:

    If your fat, sat. fat, and cholesterol intake is too high—look for what food is contributing the highest amounts of these components. This may be done using the "single nutrient" option, which shows which foods are highest in the nutrient you select—maybe sausage/pepperoni pizza? Meat-based entrée? Cream-based sauce? Whole milk? Cheese?

    Make a suggestion for a *comparable* meal item (one that you are likely to eat) with less of these nutrients—maybe thin crust, single-item pizza with less cheese? Vegetarian entrée? A cup of ice cream instead of a pint?

    Do not suggest, however, that instead of eating the 4 large slices of deep dish meat-lover's pizza and a pitcher of beer you would substitute a scoop of fat-free cottage cheese and a carrot. Reality should be taken into account.

- Make sure you mention any supplements that you take (vitamins, minerals, others), and the difference these might make to your overall nutritional profile.

- Mention other lifestyle factors (presence or lack of exercise, family history of disease, smoking, alcohol) that, along with your diet, may impact your health by affecting your risk of disease.

- If you had trouble finding the foods that you consumed in the database, mention how this may affect the accuracy of your analysis.

The following should be turned in on (or before) the due date. Please staple all pages together **in this order**, as it makes it much easier to read:

1. Your written report, addressing the questions on this handout.

2. Your personal profile page.

3. Your 3-day average information with the **nutrient summary on top.**

4. Your printouts for the three days.

TOPICAL GUIDE TO DIET ANALYSIS MODULES A THROUGH E

The diet analysis modules in this workbook each contain several exercises with worksheets. This guide is provided to allow the reader to quickly identify different exercises pertaining to specific topical areas, and then compare them to choose the individual exercises most appropriate for his or her course.

Topic	Activity(ies) or Exercise(s) Covering Topic for Each Module				
	A	B	C	D	E
Influences on food choices	1				1
Energy nutrient/alcohol intake	3		A, B	2	
Carbohydrates	6	1	A, B	3	3
Lipids	7	2	A, B	4	4
Protein	8	3, 4	A, B	5	5
Vitamins	9	5, 6	A, B	7	8
Minerals	10	7, 8	A, B	8	9
Fitness				6	10

Modules:
    Module A: Group Diet Analysis Activities
    Module B: Diet Analysis Exercises
    Module C: Diet Analysis Plus Online Assignment
    Module D: Diet Self-Study Exercises
    Module E: Thinking Through My Diet Exercises

MODULE A — GROUP DIET ANALYSIS ACTIVITIES

Author: Lorrie Miller Kohler

DIET ANALYSIS ACTIVITY DA-1: FOOD BEHAVIOR

Lesson Summary:
The purpose of this exercise is to provide a beginning of the semester atmospheric and an introductory cooperative learning group activity for students. Students will create a list of factors that influence their food choices.

Instructional Objectives:
Students will
(1)   meet and work with two other class members.
(2)   become aware of the factors which influence their food choices.
(3)   create list of reasons for food choices.

Materials:	Assignment sheet: one per group
Time required:	20 minutes

Decisions:
Group size: Three

Assignment to groups: Random assignment (possibly counting off)

Roles: Reader, Recorder, Checker

Reader: Group member who ate something most recently <u>reads</u> the problem out loud, makes certain everyone understands what the group is to do, and encourages all to participate.

Recorder: The group member on reader's left is the <u>recorder</u> who carefully records the best answers of the group members on the group answer sheet, edits what the group has written, gets the group members to check and sign the paper, then turns it in to the instructor.

Checker: the group member on reader's right is the <u>checker</u> who checks on the comprehension or learning of group members by asking them to explain or summarize material learned or discussed. S/he makes sure that everyone understands. S/he sees if all members agree before the recorder records the answer. He lets the group know how much time is left and keeps the group on task so that they finish in the allotted time.

Lesson:
Instructional task:
- Introduce the exercise by stating that the purpose of the exercise is to create a list of factors that affect food choices.
- Form groups of three randomly. Assign roles to group members.
- Review procedure for the cooperative exercise.
    o   One list from each group.
    o   Make sure everyone participates.
    o   Assist everyone in learning.
    o   Groups are not competing.

Author: Lorrie Miller Kohler   52

- Hand out the exercise (one sheet per group). Allow 15 minutes for group to create list.

Positive interdependence: One list from each group. Each group member signs final list.

Individual accountability: Randomly select one person from each group to present one reason for food choices to create on list for class.

Expected behaviors: Everyone participates in the discussion and fulfills their role. Each group member can list reasons for food choices.

Monitoring and processing:
Monitoring: Circulate among groups to check that roles are being followed and to answer questions.

Intervening: Remind groups that all members are expected to participate and know about food choice influence.

Processing: Remind groups that every member has two functions: complete the task and work as a group. Ask groups to discuss effectiveness by individually listing things that went well and things that need to be worked on.

DIET ANALYSIS ACTIVITY DA-2: TAGS—IDENTIFYING QUESTIONABLE NUTRITIONAL CLAIMS

Lesson summary:
In this exercise students will look critically at the nutritional claims made in advertising and identify those claims which are questionable.

Instructional Objectives:
The student will
(1)   become aware of nutrition claims made in advertising.
(2)   identify why a nutrition claim is questionable.
(3)   help other group members identify why a nutrition claim is questionable.

Materials: Assignment sheet: one per student

Time required: 20 minutes

Decision:
Group size: Three

Assignment to groups: Random assignment

Roles: Reader, Recorder, Checker.

Reader: Group member with the largest ad <u>reads</u> the problem, makes certain everyone understands what the group is to do, and encourages all to participate.

Recorder: The group member on the reader's left <u>records</u> the tags by putting the checks on the list.

Author: Lorrie Miller Kohler

Checker: The group member on the reader's right <u>checks</u> the answer, and keeps the group on task so that they finish in the allotted time.

Roles rotate as each ad is considered.

Lesson:

Instructional task:
- Introduce the exercise by stating the purpose of the exercise.
- Form groups of three randomly. Assign roles.
- Hand out exercise, one sheet per student.
- Review procedure for the exercise.

Positive interdependence: One ad is considered at a time by group.

Expected behaviors: Everyone participates in the discussion and fulfills their roles. Each group member is able to identify questionable nutritional claims.

DIET ANALYSIS ACTIVITY DA-3: PERCENT OF CALORIES EXERCISE

Lesson Summary:
In this exercise students will learn to determine the number of calories derived from classes of nutrients and the percent of calories from each source. They will compare the percent with the Dietary Goals for the United States.

Instructional Objectives:
Students will
(1)   calculate number of calories derived from carbohydrate, fat, and protein.
(2)   calculate the percent of calories derived from carbohydrate, fat and protein.
(3)   compare the percent of calories derived from nutrients with the respective Dietary Goals for the United States.

Materials:	Assignment sheet: one per pair during class
Time required:	30 minutes

Decisions:

Group size:	Two
Assignment to groups:	Random assignment
Roles:	Reader, Recorder

Reader: Group member who likes math the most <u>reads</u> the problem, makes certain the other person understands what is to be done.

Recorder: The other group member <u>records</u> the calculations, and keeps the pair on task so that they finish in the allotted time.

Lesson:

Instructional task:
- Introduce the exercise by stating the purpose of the exercise.
- Form groups of two randomly. Assign roles.
- Hand out exercise, one sheet per group.
- Review procedure for the exercise.
- Allow 25 minutes to complete the task.

Positive interdependence: One assignment sheet per pair.

Individual accountability: Randomly select one person from each group to present one reason for food choices to create on list for class.

Expected behaviors: Everyone participates in the discussion and fulfills their role. Each person is able to do the calculations.

DIET ANALYSIS ACTIVITY DA-4: DRI GROUP ACTIVITY

Lesson Summary:
In this exercise students will practice interpreting the Dietary Reference Intake Tables, in particular how to identify their own DRI for a specific essential nutrient. They will also identify which nutrients have RDAs, AIs and ULs, and learn how the Estimated Energy Requirement is determined.

Instructional Objectives:
Students will
(1)   identify the categories used in the DRI Tables.
(2)   identify the energy nutrients, vitamins and minerals for which there are DRIs, and the minerals for which there are ULs.
(3)   describe the method used to determine the Estimated Energy Requirement.
(4)   be familiar with the use of the tables to find individual DRIs.

Materials:	Assignment sheets: one per student, to be distributed the class period before the group exercise is scheduled; plus, one per group of three during the exercise
Time required:	30 minutes

Decisions:

Group size:	Three
Assignment to groups:	Random assignment
Roles:	Reader/Recorder, Checker, Timer.

Reader/Recorder: Reads the questions to the group and records the group's agreed upon answers on one paper.

Author: Lorrie Miller Kohler

Checker: Checks each group member's comprehension status to determine that all group members understand each question.

Timer: Keeps the group on task so as to finish in the allotted time.

Lesson:
Instructional task:
- Distribute the exercise to each student during the class period prior to the exercise.
- Introduce the exercise by stating the purpose of the exercise.
- Form groups of three randomly. Assign roles.
- Hand out exercise, one sheet per group.
- Review procedure for the exercise.
- Allow 30 minutes to complete the task.

Positive interdependence: One assignment sheet per group of three.

Expected behaviors: Everyone prepares their answers to the questions on the handout prior to class and then participates in the discussion, fulfilling their role, during class. Each person is able to use the DRI Tables and identify nutrients included in them.

DIET ANALYSIS ACTIVITY DA-5: FOOD LABEL EXERCISE

Lesson Summary:
In this exercise students will learn how to read food labels. They will read the Nutrition Facts panel and the Ingredients List on a food package. They will learn how this information can be used in selecting food.

Instructional Objectives:
Students will learn
(1)   what information is required on a food label.
(2)   what information is included in the Nutrition Facts panel of a food label and how it is organized.
(3)   what information is provided by a food package Ingredients List and how it is organized.
(4)   how this information can be used in a meaningful way in food selection.

Materials:	Food packages brought by students.
	Assignment sheet: one per student.
Time required:	20 minutes

Decisions:
Group size:	Two
Assignment to groups:	Random assignment
Roles:	Reader, Recorder

Reader: Group member who brought package <u>reads</u> the problem, makes certain the other person understands what is to be done.

Recorder: The other group member <u>records</u> the information, and keeps the pair on task so that they finish in the allotted time. Roles reverse for second package.

Lesson:

Instructional task:

- Introduce the exercise by stating the purpose of the exercise.
- Form groups of two randomly. Assign roles.
- Hand out exercise, one sheet per student.
- Review procedure for the exercise.
- Allow 10 minutes per package to complete the task.

Positive interdependence: One assignment sheet per pair.

Expected behaviors: Everyone participates in the discussion and fulfills their role. Each person is able to find information on a food package.

DIET ANALYSIS ACTIVITY DA-6:   CARBOHYDRATE AND FIBER EXERCISE
DIET ANALYSIS ACTIVITY DA-7:   LIPID EXERCISE
DIET ANALYSIS ACTIVITY DA-8:   PROTEIN EXERCISE
DIET ANALYSIS ACTIVITY DA-9:   VITAMIN EXERCISE
DIET ANALYSIS ACTIVITY DA-10:   MINERAL EXERCISE

Lesson Summary:

These five exercises may be completed using either a diet analysis of a sample student (provided as a handout) or the student's own diet analysis for a single day. As a group, students study the nutrient classes: carbohydrates, lipids, protein, vitamins, and minerals. These exercises ask the student to analyze, evaluate, and make recommendations regarding their own/the sample student's dietary intake. The activity can serve to prepare students for the evaluation of their own three-day diet.

Instructional Objectives:

Students will learn to

(1)   identify foods high in specific nutrients.
(2)   use guidelines and recommendations to evaluate a menu (or diet).
(3)   suggest menu changes that would improve the day's food intake.
(4)   calculate percent calories from carbohydrates, lipids, and proteins.

Materials: Assignment sheet for each nutrient class (use Handouts DA-6I to DA-10I if students are using their own 1-day intake reports, or Handouts DA-6S to DA10S if they are using the S. Student sample reports); computer analysis for one day's food intake (either the sample analysis or each student's own).

Time required: Varies with exercise (from 20-50 minutes)

Author: Lorrie Miller Kohler

Decisions:

Group size: Two or Three

Assignment to groups: Random assignment

Roles: Reader, Recorder, Timer

Reader: This group member <u>reads</u> the problem out loud, is responsible for getting information from the computer printout, and makes certain everyone is aware from where the information is obtained.

Recorder: The group member on the reader's right is the <u>recorder</u> who is responsible for writing the information from the computer printout. S/he also checks on the comprehension or learning of group members by asking them to explain or summarize material learned or discussed. S/he makes sure that everyone understands.

Timer: The timer is at the reader's left. S/he lets the group know how much time is left and keeps the group on task so that they finish in the allotted time. S/he is responsible for handing in any work from the group.

Lesson:

Instructional task:
- Introduce the exercise by stating what the purpose of the exercise is.
- Form groups of two or three. Assign roles to group members.
- Review procedure for the cooperative exercise.
    - Make sure everyone participates.
    - Assist everyone in learning.
    - Groups are not competing.
- Hand out exercise, one sheet per person.
- Announce amount of time allotted.

Positive interdependence: A mixture of individual and group tasks. Only one computer data sheet per group (if the provided sample student printout is used).

Expected behaviors: Everyone participates in the discussion and fulfills their role. There will be a sharing of ideas and learning from each other.

Monitoring and processing:

Monitoring: Circulate among groups to check that roles are being followed and to answer questions.

Intervening: Remind groups that all members are expected to participate and to help each other.

Processing: Remind groups that every member has two functions: complete the task and work as a group. Ask groups to discuss effectiveness by individually listing things that went well and things that need to be worked on.

Author: Lorrie Miller Kohler   58

Group members' signatures:

_____

_____

_____

DIET ANALYSIS ACTIVITY DA-1: FOOD BEHAVIOR

Objectives:
- To meet and work with two other class members
- To become aware of the factors which influence food choices
- To create a list of reasons for food choices

Instructions:

1.   Individually (on another sheet of paper) write down 3 of your most favorite foods and 3 foods you intensely dislike. (Take only 2 minutes.)

You have fifteen minutes to complete the remainder of this activity.

2.   In your group, compare your likes and dislikes, and determine what caused your acceptance/nonacceptance of these foods.

3.   Your group will make a list of the **reasons** for which you made your food choices. Choose a recorder for your group. The recorder will record the **reasons** for the likes and dislikes in the Table I found back side of this sheet.

4.   After reasons for the likes and dislikes have been recorded, each person sign off on the front side of this sheet. The recorder hands in this sheet to the instructor.

Author: Lorrie Miller Kohler

Table I: Reasons for Food Choices: Likes and Dislikes

Person	Reason for Likes	Reason for Dislikes
Person:	1. 2. 3.	1. 2. 3.
Person:	1. 2. 3.	1. 2. 3.
Person:	1. 2. 3.	1. 2. 3.

*After recording data, all group members remember to sign off on front of sheet

Name_____

Diet Analysis Activity DA-2: Tags—Identifying Questionable Nutritional Claims

Objectives:
- To look critically at the nutritional claims made in advertising.
- To become able to identify why a nutrition claim is questionable.
- To help other group members identify why a nutrition claim is questionable.

Instructions:

Each student was instructed to bring to class today an advertisement which made a nutrition claim, or a written statement of an ad from radio or television.

1.   Each of you identify and briefly write the nutritional information given in your ad.

2.   Write a statement about how accurate you believe this information is and defend your statement.

Author: Lorrie Miller Kohler

3. In groups of three, look at the ads brought by the group members, one at a time. For each, determine and agree which of the tags in the list below could be tied to each ad by checking it off on the list. The more tags you can tie on the ad, the less likely the claims are valid. Use no more than 5 minutes per ad.

TAGS

☐ Logic without proof   ☐ Misuse of terms
☐ Amateur diagnosis   ☐ Scare tactics
☐ Sales pitch   ☐ Motive: Personal gain
☐ Magical thinking   ☐ Bent truth
☐ Authority not cited   ☐ Unreliable publication
☐ Fake credentials   ☐ Incomplete truth

TAGS

☐ Logic without proof   ☐ Misuse of terms
☐ Amateur diagnosis   ☐ Scare tactics
☐ Sales pitch   ☐ Motive: Personal gain
☐ Magical thinking   ☐ Bent truth
☐ Authority not cited   ☐ Unreliable publication
☐ Fake credentials   ☐ Incomplete truth

TAGS

☐ Logic without proof   ☐ Misuse of terms
☐ Amateur diagnosis   ☐ Scare tactics
☐ Sales pitch   ☐ Motive: Personal gain
☐ Magical thinking   ☐ Bent truth
☐ Authority not cited   ☐ Unreliable publication
☐ Fake credentials   ☐ Incomplete truth

4. Each person staple his ad to his assignment sheet. As a group determine which ad made the most unfounded claims. Staple all sheets of the group together with this one on the top. Hand in.

Extension exercise: If you finish before the time is up, join another group to see what nutritional claims were made in the ads they brought.

Author: Lorrie Miller Kohler

Name_____

Name_____

DIET ANALYSIS ACTIVITY DA-3: PERCENT OF CALORIES EXERCISE

Objectives:
- To calculate number of calories derived from carbohydrate, fat, and protein.
- To calculate the percent of calories derived from carbohydrate, fat and protein.
- To compare the percent of calories derived from nutrients with the respective Dietary Goals for the United States.

Instructions:
The Nutrition Facts on a box of granola cereal shows that a 1-oz. serving (with no milk) provides the following:

calories	110
protein	2 grams
carbohydrate	22 grams
fat	2 grams

Answer the following questions. Use a calculator to help, but <u>show what calculations you have done</u>.

You will work in pairs. Be certain that both of you understand what to do and participate. Be certain that the exercise is done correctly. Watch the time. Each group hands in only one set of answers. Be sure that both names are on the answer sheet. Be sure both group members understand how to do these calculations and comparisons. Be prepared to discuss the questions.

1.   How many calories in this cereal come from:

    a. carbohydrate?   b. fat?   c. protein?

Author: Lorrie Miller Kohler

2.   Using 110 as the total number of calories in a 1-oz. serving, calculate what percent of calories in this cereal come from:

a. carbohydrate   b. fat   c. protein

3.   How does the percent of calories from each nutrient calculated in Question 2 <u>compare</u> with the U.S. Dietary Goals for:

a. carbohydrate?

b. fat?

c. protein?

Author: Lorrie Miller Kohler   64

Names:_____

_____

_____

DIET ANALYSIS ACTIVITY DA-4: DRI GROUP ACTIVITY

Directions:
Each person answer the following questions about the Dietary Reference Intakes (DRI) on this sheet. Refer to your textbook for information on essential nutrients and nutrient recommendations. You will receive 5 points for answering these questions before the next class meeting. At the next class meeting, you will work in groups of 3 to share your answers and to determine the best answer to each of these questions. At that time, your instructor will provide each group a single copy of this assignment sheet. One person in the group will record the group's agreed upon answers on the assignment sheet. You will have 30 minutes to complete your answers. Each group is responsible for turning in one paper with the group's agreed upon answers and three signatures on it. Your signature on the paper will earn you an additional 5 points for participating in the group process. Staple your completed individual assignments to the single assignment sheet completed by the group; turn all in at the end of the class.

Objectives:
Students will learn:
- how to interpret the Dietary Reference Intake Tables including categories and essential nutrients found there.
- what essential nutrients have RDAs, AIs, ULs.
- how to determine what student's specific DRI is for a particular essential nutrient.
- how the Estimated Energy Requirement is determined.

Designate the following roles in the group:

Reader/Recorder: Reads the questions to the group and records the group's agreed upon answers on one paper.

Checker: Checks each group member's comprehension status to determine that all group members understand each question.

Timer: Keeps the group on task so as to finish in the allotted time.

1. Determine the **categories** (e.g., Infants) one should look at when trying to determine the recommended nutrient intake for a particular nutrient for a given individual. List them here.

Author: Lorrie Miller Kohler

2.   List the **energy nutrients** for which there is an RDA.

3.   List the **fat-soluble vitamins** for which there is a DRI (RDA or AI).

4.   List the **water-soluble vitamins** for which there is a DRI (RDA or AI).

5.   List the **major minerals** for which there is a DRI (RDA or AI).

6.   What is done with the **major minerals** for which there is no DRI (RDA or AI)? Which are they?

7.   List the **trace minerals** for which there is a DRI (RDA or AI).

8.   For which **major minerals** and **trace minerals** is there a **UL**? List them below.

Author: Lorrie Miller Kohler

9.   Describe how the **Estimated Energy Requirement** for energy is determined.

10.   Each person in the group select an essential nutrient and determine <u>your</u> recommended intake for the nutrient. In the space below, list the nutrients and the DRI (RDA or AI) for the essential nutrient each group member has chosen.

	Essential Nutrient	**DRI (RDA or AI)**
1.		
2.		
3.		

Name of person who brought food label_____

Name of second group member   _____

Diet Analysis Activity DA-5: Food Label Exercise

Objectives:

- To read food labels.
- To learn what information is required on a food label.
- To learn what information is included in the Nutrition Facts panel of a food label and how it is organized.
- To learn what information is provided by a food package Ingredients List and how it is organized.
- To learn how this information can be used in a meaningful way in food selection.

Instructions:

Each student is to bring a food package to class. The package should have both an Ingredients List and a Nutrition Facts panel on it. Work in groups of two.

Consider one package. The person who brought the package will suggest the answers to the questions. The pair will discuss the answers and decide on what the second person will write on the assignment sheet. Hand in the sheet for the food package with the package (or the appropriate part of it) attached. Fill out the assignment sheet for the second package in a similar way.

Food Label: Items 1, 2, and 3 are three of the items required by law to appear on a food label. List them for your food package.

1.   The common name of the product. _____

2.   The name and address of the manufacturer, packer, or distributor.

3.   The net contents in terms of weight, measure, or count. _____

4.   What nutrient content claims, if any, are made on the front of the package?

5.   What health claims, if any, are made regarding the product?

Ingredient List

6.   List the ingredient that is present in the greatest proportion by weight.

Author: Lorrie Miller Kohler

7.   List the ingredient that is present in the second greatest proportion by weight.

Nutrition Facts panel

8.   List the following nutrition information for the product:

Serving size: _____

Servings per container: _____

Calories per serving: _____

Nutrition Facts	**Amount per serving**	**% Daily Value***
Total fat(g)	_____	_____
Saturated Fat(g)	_____	_____
Cholesterol(mg)	_____	_____
Sodium(mg)	_____	_____
Total carbohydrate(g)	_____	_____
Dietary fiber(g)	_____	_____
Sugars(g)	_____	_____
Protein(g)	_____	_____
Vitamin A	_____	_____
Vitamin C	_____	_____
Calcium	_____	_____
Iron	_____	_____

*Based on a 2000 calorie diet

9.   List the % Daily Values that appear to be high or low for this product.

10.   In what ways can the information from a food label help you eat a better diet?

FORMAL COOPERATIVE LEARNING GROUP ACTIVITIES

This semester you will have the opportunity to work in Formal Cooperative Learning Groups. In these groups you will work with a group of people in what is called a base group, have a specific role in the group, and follow a particular group process. The ultimate goal of formal cooperative group learning and process is to ensure the learning of each member of the group.

Early on in the semester each student will be assigned to a cooperative learning base group with three other students with whom they will work for the remainder of the semester. The task of the base group is to discuss the group activities 6S-10S. A simulated printout of a one-day dietary intake for S. Student is to be used as the basis for answering questions found in the activities on carbohydrates, lipids, proteins, vitamins and minerals. Your instructor will provide due dates for when the activities are to be completed in class. On those dates, the questions in each activity will be discussed in the assigned base groups. Since the intent of cooperative group learning is to foster the group's learning, it is important that <u>all</u> members take seriously their responsibility to complete the activities. After the group discussion is finished, each group will complete a short written evaluation on how their group functioned. Then, as a class, we will summarize the answers to the questions that comprise the activity. Since the intent of the group process is that all members of the group learn the content of a particular activity, all should be prepared to answer any question if asked to do so. All group members will turn in the particular group activity and their group's evaluation form in the assigned group's folder after the discussion is completed.

Guidelines for Cooperative Groups:
The following guidelines are important to follow when working in cooperative learning groups:

1. Learn the names of your group members.

2. When your group meets, arrange chairs so you are facing each other; everyone should be able to hear and see all members and feel included.

3. Each person should have a role in the group. Determine roles before you begin the discussion of the activity. Rotate roles at each group meeting.

4. Each student is responsible to him/herself and to his/her team.

5. Each student has an obligation to learn the material and to try to ensure that all teammates learn it.

Social Skills for Cooperative Groups:
The following social skills are important to effectively facilitate the cooperative group process:

1. Tolerant Listening

2. Constructive Disagreement

3. Asking for Clarification

4. Expressing Need for Assistance

5. Summarizing

Author: Lorrie Miller Kohler

Roles and Tasks:

The following roles and tasks will be used in our cooperative groups:

1. <u>Reader</u>: reads aloud the questions in the exercise to the group members.

2. <u>Checker</u>: asks each person to state their answer to the questions and asks if all agree/disagree. <u>All group members should agree on answer to question and understand answer.</u>

3. <u>Summarizer</u>: summarizes the answer to each question after discussion of the question is completed by the group.

4. <u>Timer</u>: watches the time, keeps group on task and makes sure group finishes in allotted time. At the completion of the exercise, <u>timer</u> sees that all papers are placed in the group's folder and that <u>ABSENTEES</u> are recorded on the attendance sheet inside the folder. Folders are turned in to instructor.

Name_____

DIET ANALYSIS ACTIVITY DA-6S: CARBOHYDRATE AND FIBER EXERCISE

Objectives:
- To identify foods high in simple and complex carbohydrates and in fiber.
- To use guidelines and recommendations to evaluate a menu (or diet).
- To suggest menu changes that would improve carbohydrate and fiber in the diet.
- To calculate percent calories from carbohydrates.

Instructions:
Answer the following questions about carbohydrates, fiber and calories using the simulated computer printout for S. Student which you have been given.

1. Using the Spreadsheet for Day 1, list the foods eaten by S. Student in order from highest to lowest according to amount (in grams) of carbohydrate they contain. Also record the quantity of food. Do the same for fiber containing foods.

Carbohydrate-rich foods (more than 5 grams)		Fiber-rich foods (more than 1 gram)	

Author: Lorrie Miller Kohler

2.   Which foods contain little or no carbohydrate? fiber?

Little/no carbohydrate (less than 1 gram)		Little/no fiber (less than 1 gram)	

3.   a.   Are the foods listed in #1 which are high in carbohydrate and fiber of plant or animal
                origin?

        b.   Are the foods listed in #2 of plant or animal origin?

        c.   From questions 3a and b, what do you conclude about the sources carbohydrates and fiber?

4.   Which foods contain complex carbohydrates (starch)? simple carbohydrates?

Complex carbohydrates (starch)	Simple carbohydrates

5.   **Each group member** suggest a change you would make to increase the complex carbohydrates (starch) in this day's intake. Record suggestions below.

6.   Would you make any changes in the fiber intake? If so, describe one. If not, why not? Record group members' changes or reasons for not changing below.

7.   Calculate the percent of calories in the doughnut that came from carbohydrate. Be sure **all** group members understand how to do these calculations. Show your math.

8.   Calculate what percent of the total calories came from carbohydrate. Be sure **all** group members understand how to do these calculations. Show your math.

Author: Lorrie Miller Kohler

9.   In a sentence, describe how the percent of total calories from carbohydrate in this day's intake compares with the Dietary Goals.

10.   Do you think this day's intake implemented the Dietary Guidelines for Americans related to calories, carbohydrates and fiber? (Review Dietary Guidelines.) If so, how? If not, how not? Record group members' responses below.

All Reports - Day 1

Profile Info

Name	S. Student
Birthdate	Mar 23, 1977
Gender	Female
Height	5 ft 6 in
Weight	130 lb
Activity Level	Lightly Active
BMI	20.98

Recommendations

Basic Components

Calories	2028.62
Protein	47.17 g
Carbohydrates	278.93 g
Dietary Fiber	28.4 g
Fat - Total	63.11 g
Saturated Fat	20.29 g
Mono Fat	22.54 g
Poly Fat	20.29 g
Cholesterol	300 mg

Vitamins

Vitamin A RE	700 RE
Thiamin-B1	1.1 mg
Riboflavin-B2	1.1 mg
Niacin-B3	14 mg
Vitamin-B6	1.3 mg
Vitamin-B12	2.4 mcg
Vitamin C	75 mg
Vitamin D mcg	5 mcg
Vit E-Alpha Equiv.	15 mg
Folate	400 mcg

Minerals

Calcium	1000 mg
Iron	18 mg
Magnesium	310 mg
Phosphorus	700 mg
Potassium	3500 mg
Sodium	2400 mg
Zinc	8 mg

Author: Lorrie Miller Kohler

Activity

No activities for this day.

Intake

Item	Meal	Amount
Juice, orange, unswtnd, prep f/fzn conc w/water	Breakfast	0.5 c
Cereal, corn flakes, rte, dry (Kellogg's Company)	Breakfast	1.0 c
Milk, nonfat/skim, w/add vit A	Breakfast	0.5 c
Beef, ground, hamburger patty, brld, well done, 18% fat	Lunch	4.0 oz-wt
Buns, hamburger	Lunch	1.0 ea
Milk, nonfat/skim, w/add vit A	Lunch	1.0 c
Catsup/Ketchup	Lunch	1.0 tbs
Doughnut, cake, med, 3 1/4" diameter	Lunch	1.0 ea
Pork, chop, center loin, pan fried	Dinner	3.0 oz-wt
Potatoes, mashed, w/whole milk	Dinner	1.0 c
Cabbage, raw, shredded, cup	Dinner	0.5 c
Mayonnaise, soybean oil, w/salt	Dinner	1.0 tbs
Corn, yellow, sweet, kernels, vac pack, cnd, cup	Dinner	0.5 c
Butter, salted, cup	Dinner	1.0 tbs
Pudding, chocolate, prep f/dry mix w/whole milk	Dinner	1.0 c
Coffee, brewed, prep w/tap water	Dinner	2.0 c
Crackers, standard, reg, snack type, round	Snack	4.0 ea
Peanut Butter, creamy, w/salt	Snack	1.5 tbs

Author: Lorrie Miller Kohler

78

Bar Graph

Weight: 1995.25 g   **Water**: 78%

Nutrient	Value	Goal %	0% 50 100 150%
Basic Components			
Calories	2156.4	106%	
Calories from Fat	873.68	--%	
Calories from Saturated Fat	305.7	--%	
Protein	102.66 g	218%	
Carbohydrates	228.99 g	82%	
Dietary Fiber	14.63 g	52%	
Fat - Total	97.08 g	154%	
Saturated Fat	33.97 g	167%	
Mono Fat	36.46 g	162%	
Poly Fat	18.62 g	92%	
Cholesterol	294.58 mg	98%	
Water	1548.05 g	--%	
Vitamins			
Vitamin A RE	702.74 RE	100%	
Thiamin-B1	2.29 mg	208%	
Riboflavin-B2	2.4 mg	218%	
Niacin-B3	28.2 mg	201%	
Vitamin-B6	2.28 mg	175%	
Vitamin-B12	5.97 mcg	249%	
Vitamin C	105.18 mg	140%	
Vitamin D mcg	8.37 mcg	167%	
Vit E-Alpha Equiv.	7.39 mg	49%	
Folate	377.93 mcg	94%	
Minerals			
Calcium	1014.59 mg	101%	
Iron	18.24 mg	101%	
Magnesium	304.72 mg	98%	
Phosphorus	1566.7 mg	224%	
Potassium	3634.79 mg	104%	
Sodium	2972.48 mg	124%	
Zinc	14.51 mg	181%	
Others			
Alcohol	0 g	--%	
Caffeine	280.6 mg	--%	

Author: Lorrie Miller Kohler

Spreadsheet

Name	Amount	Weight (g)	Cals	Cal From Fat	Cals Sat Fat	Prot (g)	Carb (g)
Juice, orange, unswtnd, pre...	0.5 c	124.5	56.02	0.63	0.04	0.84	13.42
Cereal, corn flakes, rte, d...	1.0 c	28	102.2	1.71	0.45	1.83	24.21
Milk, nonfat/skim, w/add vi...	0.5 c	122.5	42.88	1.98	1.26	4.18	5.94
Beef, ground, hamburger pat...	4.0 oz-wt	113.4	317.51	180.0	70.68	31.97	0.0
Buns, hamburger	1.0 ea	43	122.98	19.71	4.59	3.65	21.62
Milk, nonfat/skim, w/add vi...	1.0 c	245	85.75	3.96	2.52	8.35	11.88
Catsup/Ketchup	1.0 tbs	15	15.6	0.48	0.06	0.23	4.09
Doughnut	1.0 ea	47	197.87	96.84	15.3	2.35	23.35
Pork, chop, center loin, pa...	3.0 oz-wt	85.05	235.58	126.69	45.93	25.43	0.0
Potatoes, mashed, w/whole m...	1.0 c	210	161.7	11.07	6.21	4.07	36.85
Cabbage, raw, shredded, cup	0.5 c	35	8.75	0.81	0.09	0.5	1.9
Mayonnaise, soybean oil, w/...	1.0 tbs	13.8	98.94	98.55	14.58	0.15	0.37
Corn, yellow, sweet, kernel...	0.5 c	105	82.95	4.73	0.72	2.53	20.41
Butter, salted, cup	1.0 tbs	14	100.38	102.15	63.54	0.11	0.01
Pudding, chocolate, prep f/...	1.0 c	284	315.24	86.76	53.28	9.08	51.12
Coffee, brewed, prep w/tap ...	2.0 c	474	9.48	0.13	0.04	0.46	1.88
Crackers, standard, reg, sn...	4.0 ea	12	60.24	27.32	4.07	0.89	7.32
Peanut Butter, creamy, w/sa...	1.5 tbs	24	142.32	110.16	22.34	6.04	4.62
Total		1995.25	2156.4	873.68	305.7	102.66	228.99

Name	Fiber (g)	Fat-T (g)	Fat-S (g)	Fat-M (g)	Fat-P (g)	Chol (mg)	H2O (g)
Juice, orange, unswtnd, pre...	0.24	0.07	0.0	0.01	0.01	0.0	109.68
Cereal, corn flakes, rte, d...	0.78	0.19	0.05	0.02	0.11	0.0	0.9
Milk, nonfat/skim, w/add vi...	0.0	0.22	0.14	0.06	0.0	2.45	111.23
Beef, ground, hamburger pat...	0.0	20.0	7.85	8.75	0.75	114.53	59.93
Buns, hamburger	1.16	2.19	0.51	0.36	1.08	0.0	14.62
Milk, nonfat/skim, w/add vi...	0.0	0.44	0.28	0.11	0.01	4.9	222.46
Catsup/Ketchup	0.2	0.05	0.01	0.01	0.02	0.0	9.99
Doughnut	0.7	10.76	1.7	4.36	3.7	17.39	9.77
Pork, chop, center loin, pa...	0.0	14.08	5.1	6.0	1.61	78.23	45.05
Potatoes, mashed, w/whole m...	4.2	1.23	0.69	0.3	0.11	4.2	164.76
Cabbage, raw, shredded, cup	0.8	0.09	0.01	0.0	0.04	0.0	32.25
Mayonnaise, soybean oil, w/...	0.0	10.95	1.62	3.13	5.69	8.14	2.11
Corn, yellow, sweet, kernel...	2.1	0.52	0.08	0.15	0.24	0.0	80.4
Butter, salted, cup	0.0	11.35	7.06	3.27	0.42	30.66	2.22
Pudding, chocolate, prep f/...	2.84	9.64	5.92	2.84	0.36	34.08	211.28
Coffee, brewed, prep w/tap ...	0.0	0.01	0.0	0.0	0.01	0.0	470.68
Crackers, standard, reg, sn...	0.19	3.04	0.45	1.28	1.14	0.0	0.42
Peanut Butter, creamy, w/sa...	1.41	12.24	2.48	5.82	3.31	0.0	0.29
Total	14.63	97.08	33.97	36.46	18.62	294.58	1548.05

Name	A-RE (RE)	B1 (mg)	B2 (mg)	B3 (mg)	B6 (mg)	B12 (mcg)	Vit C (mg)
Juice, orange, unswtnd, pre...	9.96	0.1	0.02	0.25	0.05	0.0	48.43
Cereal, corn flakes, rte, d...	210.28	0.36	0.39	4.67	0.47	0.0	14.0
Milk, nonfat/skim, w/add vi...	74.72	0.04	0.17	0.1	0.04	0.46	1.23
Beef, ground, hamburger pat...	0.0	0.07	0.27	6.76	0.33	3.08	0.0
Buns, hamburger	0.0	0.2	0.13	1.69	0.01	0.02	0.04
Milk, nonfat/skim, w/add vi...	149.45	0.08	0.34	0.21	0.09	0.93	2.45
Catsup/Ketchup	15.3	0.01	0.01	0.2	0.03	0.0	2.26
Doughnut	7.99	0.1	0.11	0.87	0.02	0.12	0.09
Pork, chop, center loin, pa...	1.7	0.96	0.25	4.75	0.39	0.61	0.85
Potatoes, mashed, w/whole m...	12.6	0.18	0.08	2.34	0.48	0.0	14.07
Cabbage, raw, shredded, cup	4.54	0.02	0.01	0.1	0.03	0.0	11.27
Mayonnaise, soybean oil, w/...	11.59	0.0	0.0	0.0	0.07	0.03	0.0
Corn, yellow, sweet, kernel...	25.2	0.04	0.08	1.23	0.06	0.0	8.51
Butter, salted, cup	105.56	0.0	0.0	0.01	0.0	0.01	0.0
Pudding, chocolate, prep f/...	73.84	0.08	0.48	0.28	0.1	0.7	1.98
Coffee, brewed, prep w/tap ...	0.0	0.0	0.0	1.04	0.0	0.0	0.0
Crackers, standard, reg, sn...	0.0	0.05	0.04	0.48	0.0	0.0	0.0
Peanut Butter, creamy, w/sa...	0.0	0.02	0.02	3.21	0.11	0.0	0.0
Total	702.74	2.29	2.4	28.2	2.28	5.97	105.18

Name	D-mcg (mcg)	E-aTE (mg)	Fola (mcg)	Calc (mg)	Iron (mg)	Magn (mg)	Phos (mg)
Juice, orange, unswtnd, pre...	0.0	0.24	54.78	11.2	0.12	12.45	19.92
Cereal, corn flakes, rte, d...	1.0	0.03	98.84	1.12	8.68	3.36	10.92
Milk, nonfat/skim, w/add vi...	1.23	0.04	6.12	150.68	0.04	13.48	123.72
Beef, ground, hamburger pat...	0.33	0.23	12.47	13.6	2.77	27.21	206.38
Buns, hamburger	0.0	0.67	40.85	59.77	1.36	8.6	37.84
Milk, nonfat/skim, w/add vi...	2.45	0.09	12.25	301.35	0.09	26.95	247.45
Catsup/Ketchup	0.0	0.22	2.25	2.85	0.1	3.3	5.85
Doughnut	0.02	1.8	22.09	20.68	0.91	9.4	126.43
Pork, chop, center loin, pa...	0.25	0.22	5.1	22.96	0.76	24.66	220.27
Potatoes, mashed, w/whole m...	0.42	0.1	16.8	54.6	0.56	37.8	100.8
Cabbage, raw, shredded, cup	0.0	0.04	15.05	16.45	0.2	5.25	8.05
Mayonnaise, soybean oil, w/...	0.04	0.31	1.1	2.48	0.06	0.13	3.86
Corn, yellow, sweet, kernel...	0.0	0.09	51.45	5.25	0.44	24.15	67.2
Butter, salted, cup	0.19	0.22	0.42	3.36	0.02	0.28	3.22
Pudding, chocolate, prep f/...	2.44	0.16	11.36	315.24	1.02	42.6	264.12
Coffee, brewed, prep w/tap ...	0.0	0.0	0.0	9.48	0.22	23.7	4.74
Crackers, standard, reg, sn...	0.0	0.54	9.24	14.4	0.43	3.24	27.36
Peanut Butter, creamy, w/sa...	0.0	2.4	17.76	9.12	0.43	38.16	88.56
Total	8.37	7.39	377.93	1014.59	18.24	304.72	1566.7

Author: Lorrie Miller Kohler

Name	Potas (mg)	Sod (mg)	Zinc (mg)	Alco (g)	Caff (mg)
Juice, orange, unswtnd, pre...	236.55	1.25	0.06	0.0	0.0
Cereal, corn flakes, rte, d...	25.48	297.92	0.16	0.0	0.0
Milk, nonfat/skim, w/add vi...	203.35	63.7	0.49	0.0	0.0
Beef, ground, hamburger pat...	395.75	100.92	7.03	0.0	0.0
Buns, hamburger	60.63	240.8	0.26	0.0	0.0
Milk, nonfat/skim, w/add vi...	406.7	127.4	0.98	0.0	0.0
Catsup/Ketchup	72.15	177.9	0.03	0.0	0.0
Doughnut	59.69	256.62	0.25	0.0	0.0
Pork, chop, center loin, pa...	361.45	68.04	1.96	0.0	0.0
Potatoes, mashed, w/whole m...	627.9	636.3	0.6	0.0	0.0
Cabbage, raw, shredded, cup	86.1	6.3	0.06	0.0	0.0
Mayonnaise, soybean oil, w/...	4.69	78.38	0.02	0.0	0.0
Corn, yellow, sweet, kernel...	195.3	285.6	0.48	0.0	0.0
Butter, salted, cup	3.64	115.64	0.01	0.0	0.0
Pudding, chocolate, prep f/...	462.92	292.52	1.26	0.0	5.68
Coffee, brewed, prep w/tap ...	255.96	9.48	0.08	0.0	274.92
Crackers, standard, reg, sn...	15.96	101.64	0.08	0.0	0.0
Peanut Butter, creamy, w/sa...	160.56	112.08	0.7	0.0	0.0
Total	3634.79	2972.48	14.51	0.0	280.6

Pyramid

Group	Recommended Servings	Servings Consumed
Fats, Oils & Sweets	Use sparingly	14.27
Milk, Yogurt & Cheese	2 - 3	2.12
Meat, Poultry, Fish, Dry Beans, Eggs & Nuts	2 - 3	3.08
Fruits	2 - 4	0.66
Vegetables	3 - 5	3.49
Bread, Cereal, Rice & Pasta	6 - 11	3.61

Ratios

Water: 78%

Source of Calories		0%	25	50	75	100%
Protein	19%					
Carbohydrates	42%					
Fat - Total	40%					
Alcohol	0%					

Source of Fat		0%	25	50	75	100%
Saturated Fat	14%					
Mono Fat	15%					
Poly Fat	8%					
Other Fat	3%					

Exchanges

Starch	6.28	Fruit	0.9
Other Carbs	5.18	Vegetables	0.35
Very Lean Meat	--	Milk	1.48
Meat	8.79		

Ratios

P:S (Poly / Saturated Fat)	0.55 : 1
Potassium : Sodium	1.22 : 1
Calcium : Phosphorus	0.65 : 1
CSI (Chol. / Sat Fat Index)	49.04

Nutrition Facts

Serving Size (1995g)

Amount Per Serving

Calories 2156	Calories from Fat 874

% Daily Value

	% Daily Value
Fat - Total 97g	**154**%
Saturated Fat 34g	**167**%
Cholesterol 295mg	**98**%
Sodium 2972mg	**124**%
Carbohydrates 229g	**82**%
Dietary Fiber 15g	**52**%
Protein 103g	**218**%

Vitamin A 100%	Vitamin C 140%
Calcium 101%	Iron 101%

*Percent Daily Values are based on your custom profile.

Version: 1.0 ESHA Research Thomson Learning

Author: Lorrie Miller Kohler

Name_____

Diet Analysis Activity DA-7S: Lipid Exercise

Objectives:
- To identify foods high in saturated and unsaturated fats, and in cholesterol.
- To calculate percent fat by weight and percent calories from fat.
- To use guidelines and recommendations to evaluate a menu for fat content.
- To suggest dietary changes that would improve fat and cholesterol intake.

Instructions:
Answer the following questions about fats, cholesterol, and calories using the simulated computer printout for S. Student which you have been given.

1. Review the food <u>Intake</u> list from the simulated computer printout. Identify and list foods <u>you</u> believe high in total fat and in cholesterol.

Foods high in total fat	Foods high in cholesterol

2. a. Using the <u>Spreadsheet for Day 1</u>, identify those foods containing 3 or more grams of total fat, saturated fat, monounsaturated fat, and polyunsaturated fat, and list them in order from high to low. Also record quantity of food.

Total fat	Saturated fat	Monounsat. fat	Polyunsat. fat

Author: Lorrie Miller Kohler

b.   Do the same for those foods containing 5 or more milligrams of cholesterol. Record milligrams of cholesterol and quantity of food.

Cholesterol			

3.   The ground beef patty weighs 113 grams, 20 grams of which are fat. Calculate % fat by weight in this ground beef patty by dividing the weight of the fat by the total weight of the patty and multiply by 100.

20 grams/113 grams =_____x 100 =_____% fat by weight in the ground beef patty

4.   Calculate % calories from fat for the ground beef patty.
Step 1. Calculate the number of calories in 20 grams of fat.
Step 2. Divide that number by total calories in the patty (318 cal.).
Step 3. Multiply by 100.

Calories from fat/total patty cal. =_____ x 100% =_____ % cal. from fat in ground beef patty

5.   Write a single sentence describing how the answer from Question #3 is different from the answer from Question #4.

6.   If the total caloric intake for this day is 2156 calories, then what percent of total calories was contributed by total fat in the beef patty?
Calculate this by:
Step 1. Multiplying 20 grams x 9 = calories from fat in beef patty.
Step 2. Dividing this number by 2156 and multiplying by 100.

Calories from fat/total calories for day x 100 = _____% calories from fat in the beef patty

7.   Note how Questions 3, 4, and 6 are asking different questions. As a group agree on and summarize the differences in a single sentence and write it below.

8.   What percent of the total calories come from saturated, monounsaturated, and polyunsaturated fat? The total grams of these fats is given in the Spreadsheet for Day 1. To calculate the amounts, multiply the number of grams of each type of fat by 9; then divide each answer by 2156 (total calories); and multiply by 100. Show your math.

_____% calories from saturated fat

_____% calories from monounsaturated fat

_____% calories from polyunsaturated fat

9.   How do the values in Question 8 compare with recommendations for total fat intake of 20%-35% of calories with less than 10% of calories coming from saturated fat? As a group summarize the comparisons in a sentence and write it here.

10.   Would you recommend any changes in the foods containing fats that were selected in this day's intake? If yes, what suggestion(s)? If not, why not? Record group's responses below.

Author: Lorrie Miller Kohler

11.   In Question #2 you listed foods containing cholesterol from in this day's intake. What is the origin—plant or animal—of these foods?

12.   What is the <u>source</u> of cholesterol in the doughnut: _____, mashed potatoes:_____, mayonnaise:_____, and pudding:_____?

13.   How does the total cholesterol consumed in this day's intake compare with the recommendation to consume less than 300 milligrams of cholesterol a day?

14.   Would you make any suggestions for change in cholesterol intake of S. Student? If yes, what suggestion(s)? If no, why not? Record changes or reasons for no change below.

Name_____

DIET ANALYSIS ACTIVITY DA-8S: PROTEIN EXERCISE

Objectives:
- To identify foods high in quantity of protein.
- To evaluate the adequacy of protein intake in grams and in the context of caloric intake.
- To compare the quality of protein in foods from plant and animal sources.
- To calculate percent calories from protein.
- To suggest menu choices that are high in protein and fiber and low in fat.

Instructions:

Answer the following questions about proteins using the simulated computer printout for S. Student which you have been given.

1. Review the foods eaten by S. Student using the Spreadsheet for Day 1 and identify the foods that contain more than 3 grams of protein. List these foods in decreasing order of amount of protein and according to their source. Also record quantity of food.

Animal-derived protein		Plant-derived protein	

89   Author: Lorrie Miller Kohler

2.   Compare the quantity of protein obtained from these two sources. As a group agree on and describe in a single sentence the comparison. Record it below.

3.   Describe and compare the total number of grams of protein (_____) with the RDA for protein (_____) for S. Student. * Remember the number of grams of protein consumed should be no more than twice the RDA for protein. What is twice the RDA for S. Student?_____ Make this comparison in the context of looking at her calorie intake (_____) for the day and her RDA for calories (_____). In looking at the gram intake in the context of the caloric intake, is the quantity of her protein intake for this day adequate? Why or why not?

4.   Each group **member** compare the quality of protein from the two sources (in Question #1). As a group describe the comparison in a single sentence. Record it below.

5.   In the space below <u>rank</u> all the protein-rich foods listed in Question #1 from <u>highest to</u> <u>lowest</u> according to the amount (in grams) of total fat. Record number of grams of fat.

6.   After looking at the data in Question #5, as a group, make a statement comparing the <u>source</u> of high protein-rich foods with their fat content.

7.   What percent of the total calories for the day came from protein? <u>Show your calculations below</u>.

Author: Lorrie Miller Kohler

8. Do you think this day's intake implemented the Dietary Goals for protein? If yes, how? If no, how not? Record group members' responses.

9. As a group delete one of this day's high-protein foods and add another protein source which **would not change the amount of protein, but would decrease the amount of total fat and increase the amount of fiber.** Use a food composition table to verify that your choice is a good one.

Name the food you are deleting. _____

- How much protein does it contain? _____

- How much fat does it contain? _____

- How much fiber does it contain? _____

Name and give the amount of the food you are adding. _____

- Indicate the amount of protein in the food you are adding _____

- Indicate the amount of fat in the food you are adding _____

- Indicate the amount of fiber in the food you are adding _____

Name_____

DIET ANALYSIS ACTIVITY DA-9S: VITAMIN EXERCISE

Objectives:
- To identify food sources of selected vitamins.
- To use RDAs to evaluate intake of selected vitamins.
- To suggest menu choices that would provide adequate intake of selected vitamins.
- To identify major functions of selected vitamins.

Instructions:
Answer the following questions about vitamins using the simulated computer printout for S. Student which you have been given.

1.   Review the foods eaten by S. Student using the Spreadsheet for Day 1 and identify the foods that contain 10 milligrams or more of vitamin C. List these foods in <u>decreasing order</u> of amount in milligrams of vitamin C they contain. Also record quantity of food.

2.   Identify and list the foods that provide no vitamin C and determine their source (animal or plant). What do you observe about the source of vitamin-containing foods?

     Author: Lorrie Miller Kohler

3.   Compare and record the number of milligrams of vitamin C in the intake of S. Student with the RDA. Is it too high, too low, or about right?

4.   As a group delete one of this day's high vitamin C foods and add another that would not significantly change the amount of vitamin C but would <u>increase</u> the <u>amount of fiber</u>. Use a food composition table to verify that your choice is a good one.

Name the food you are deleting. _____

•   Indicate the amount of vitamin C in deleted food. _____

•   Indicate the amount of fiber in the deleted food. _____

Name and give the amount of food you are adding. _____

•   Indicate the amount of Vitamin C in added food. _____

•   Indicate the amount of fiber in the added food. _____

5.   List three major functions of vitamin C in the body.

6. Identify the foods eaten by S. Student that contain 10 or more REs of vitamin A. List them in order of decreasing amounts of RE of vitamin A they contain. Also record quantity of food. If the food contains preformed vitamin A place (**PA**) after the food, if it contains beta carotene, place (**BC**) after the food, and if it has been fortified with vitamin A, place (**F**) after the food.

7. Compare and record the number of REs of vitamin A in this day's intake with the RDA for vitamin A. Is it too high, too low, or about right?

8. What is a vitamin A-rich fruit that would be a good replacement for the pudding dessert?

9. What is a vitamin A-rich vegetable that would be a good replacement for the corn?

Author: Lorrie Miller Kohler

10. List three major functions of vitamin A in the body.

11. For each of the following B vitamins, identify the two major sources in this day's diet.

thiamin (vit. B_1):

riboflavin (vit. B_2):

niacin (vit. B_3):

vitamin B_6:

vitamin B_{12}:

folate:

12. As a group make a general statement about the major function of the B vitamins.

13. Why do you think the corn flakes are high in many of the vitamins?

Name_____

DIET ANALYSIS ACTIVITY DA-10S: MINERAL EXERCISE

Objectives:
- To identify food sources of selected minerals.
- To use guidelines and recommendations to evaluate intake of selected minerals.
- To suggest menu choices that would provide adequate intake of selected minerals.
- To identify major functions of selected minerals.

Instructions:
Answer the following questions about minerals using the simulated computer printout for S. Student which you have been given.

1.   Review the foods eaten by S. Student using the Spreadsheet for Day 1 and identify the foods that contain 10 milligrams (mg.) or more of calcium. List these foods in decreasing order of amount of calcium they contain. Also record quantity of food.

Foods containing 10 mgs. or more of calcium	

2.   Compare and record the number of milligrams of calcium in this day's intake with the DRI (AI) for S. Student. Is it too high, too low, or about right?

3.   As a group add a food of <u>plant</u> origin that is a good source of calcium. Use a food composition table to verify that your choice is a good one.

Name and give the amount of the food you are adding. _____

How much calcium is there in the food you are adding? _____

Author: Lorrie Miller Kohler

4. List two major functions of calcium in the body.

5. Identify the foods eaten by S. Student that contain 0.7 milligrams (mg.) or more of iron. List them in order of decreasing amounts of milligrams of iron they contain. Also record quantity of food.

Foods containing 0.7 mgs. or more of iron		

6. Compare and record the number of milligrams of iron in this day's intake with the RDA for S. Student. Is it too high, too low, or about right?

7. As a group change this day's intake to increase the amount of iron. List 3 additional food sources of iron and the amount of iron that each contains.

    1.

    2.

    3.

8. Describe the major function of iron in the body.

9. Why do the hamburger bun, doughnut and cornflakes have relatively high iron content (as compared with other foods in this day's intake)?

10. What is the total amount of sodium consumed in this day?

Author: Lorrie Miller Kohler   98

11.   Do you think that this intake is high, low, or about right? Upon what guideline or recommendation (other than the computer printout) are you basing your opinion?

12.   Carefully review the foods and their sodium content. As a group suggest <u>three</u> specific menu changes that would lower the sodium content of this day's intake. List the <u>food</u> and the <u>amount</u> of sodium it contains.

    1.

    2.

    3.

Author: Lorrie Miller Kohler

COOPERATIVE GROUP'S EVALUATION FORM

At the completion of your Base Group's activity, complete this evaluation form. The Reader/ Recorder reads the questions to the group, and records the group's agreed upon answers. Place the Evaluation Form into the group's folder and turn it in with the completed group activity.

1.　　Overall, <u>how effectively</u> did your group work together on this assignment? (Circle the appropriate response.)

1	2	3	4	5
not at all	poorly	adequately	well	extremely well

2.　　<u>How many</u> of the total number of group members <u>participated actively</u> most of the time? (Circle the appropriate number.)

　　0　　　　　1　　　　　2　　　　　3　　　　　4

3.　　<u>How many</u> of the group's members <u>were fully prepared</u> for the group work most of the time? (Circle the appropriate number.)

　　0　　　　　1　　　　　2　　　　　3　　　　　4

4.　　Each person give one specific example of <u>something you learned from the group</u> that you probably wouldn't have learned on your own.

5.　　Suggest <u>one specific practical change</u> the group could make that would help improve everyone's learning.

6.　　Describe how well each group member carried out his or her respective role.

Author: Lorrie Miller Kohler　　　　　　100

FORMAL COOPERATIVE LEARNING GROUP ACTIVITIES

This semester you will have the opportunity to work in Formal Cooperative Learning Groups. In these groups you will work with a group of people in what is called a base group, have a specific role in the group, and follow a particular group process. The ultimate goal of formal cooperative group learning and process is to ensure the learning of each member of the group.

Early on in the semester each student will be assigned to a cooperative learning base group with three other students with whom they will work for the remainder of the semester. The task of the base group is to discuss the group activities 6I-10I. Each student's 1-day Diet Analysis Plus printout is to be used as the basis for answering questions found in the activities on carbohydrates, lipids, proteins, vitamins and minerals. Your instructor will provide due dates for when the activities are to be completed in class. On those dates, the questions in each activity will be discussed in the assigned base groups. Since the intent of cooperative group learning is to foster the group's learning, it is important that all members take seriously their responsibility to complete the activities. After the group discussion is finished, each group will complete a short written evaluation on how their group functioned. Then, as a class, we will summarize the answers to the questions that comprise the activity. Since the intent of the group process is that all members of the group learn the content of a particular activity, all should be prepared to answer any question if asked to do so. All group members will turn in the particular group activity and their group's evaluation form in the assigned group's folder after the discussion is completed.

Guidelines for Cooperative Groups:
The following guidelines are important to follow when working in cooperative learning groups:

1. Learn the names of your group members.
2. When your group meets, arrange chairs so you are facing each other; everyone should be able to hear and see all members and feel included.
3. Each person should have a role in the group. Determine roles before you begin the discussion of the activity. Rotate roles at each group meeting.
4. Each student is responsible to him/herself and to his/her team.
5. Each student has an obligation to learn the material and to try to ensure that all teammates learn it.

Social Skills for Cooperative Groups:
The following social skills are important to effectively facilitate the cooperative group process:

1. Tolerant Listening
2. Constructive Disagreement
3. Asking for Clarification
4. Expressing Need for Assistance
5. Summarizing

Author: Lorrie Miller Kohler

Roles and Tasks:

The following roles and tasks will be used in our cooperative groups:

1. <u>Reader</u>: reads aloud the questions in the exercise to the group members.

2. <u>Checker</u>: asks each person to state their answer to the questions and asks if all agree/disagree. <u>All group members should agree on answer to question and understand answer.</u>

3. <u>Summarizer</u>: summarizes the answer to each question after discussion of the question is completed by the group.

4. <u>Timer</u>: watches the time, keeps group on task and makes sure group finishes in allotted time. At the completion of the exercise, <u>timer</u> sees that all papers are placed in the group's folder and that <u>ABSENTEES</u> are recorded on the attendance sheet inside the folder. Folders are turned in to instructor.

Name_____

DIET ANALYSIS ACTIVITY DA-6I: CARBOHYDRATE AND FIBER EXERCISE

Objectives:
- To identify foods high in simple and complex carbohydrates and in fiber.
- To use guidelines and recommendations to evaluate a menu (or diet).
- To suggest menu changes that would improve carbohydrate and fiber in the diet.
- To calculate percent calories from carbohydrates.

Instructions:

Answer the following questions about carbohydrates, fiber and calories using your Diet Analysis Plus printout.

1. Using the Spreadsheet for Day 1, list the foods you ate in order from highest to lowest according to amount (in grams) of carbohydrate they contain. Also record the quantity of food. Do the same for fiber containing foods.

Carbohydrate-rich foods (more than 5 grams)		Fiber-rich foods (more than 1 gram)	

Author: Lorrie Miller Kohler

2.   Which foods contain little or no carbohydrate? fiber?

Little/no carbohydrate (less than 1 gram)		Little/no fiber (less than 1 gram)	

3.   a.   Are the foods listed in #1 which are high in carbohydrate and fiber of plant or animal origin?

        b.   Are the foods listed in #2 of plant or animal origin?

        c.   From questions 3a and b, what do you conclude about the sources carbohydrates and fiber?

4.   Which foods contain complex carbohydrates (starch)? simple carbohydrates?

Complex carbohydrates (starch)	Simple carbohydrates

5.   **Each group member** suggest a change you would make to increase the complex carbohydrates (starch) in this day's intake. Record suggestions below.

6.   Would you make any changes in the fiber intake? If so, describe one. If not, why not? Record group members' changes or reasons for not changing below.

7.   Select a food item and calculate the percent of calories in this food that came from carbohydrate. Be sure **all** group members understand how to do these calculations. Show your math.

8.   Calculate what percent of the total calories came from carbohydrate. Be sure **all** group members understand how to do these calculations. Show your math.

Author: Lorrie Miller Kohler

9.   In a sentence, describe how the percent of total calories from carbohydrate in this day's intake compares with the Dietary Goals.

10. Do you think this day's intake implemented the Dietary Guidelines for Americans related to calories, carbohydrates and fiber? (Review Dietary Guidelines if necessary.) If so, how? If not, how not? Record group members' responses below.

Name_____

DIET ANALYSIS ACTIVITY DA-7I: LIPID EXERCISE

Objectives:
- To identify foods high in saturated and unsaturated fats, and in cholesterol.
- To calculate percent fat by weight and percent calories from fat.
- To use guidelines and recommendations to evaluate a menu for fat content.
- To suggest dietary changes that would improve fat and cholesterol intake.

Instructions:
Answer the following questions about fats, cholesterol, and calories using your 1-day Diet Analysis Plus printout.

1.   Review the food <u>Intake</u> list from the diet analysis printout. Identify and list foods <u>you</u> believe high in total fat and in cholesterol.

Foods high in total fat	Foods high in cholesterol

2.   a.   Using the <u>Spreadsheet for Day 1</u>, identify those foods containing 3 or more grams of total fat, saturated fat, monounsaturated fat, and polyunsaturated fat, and list them in order from high to low. Also record quantity of food.

Total fat	Saturated fat	Monounsat. fat	Polyunsat. fat

Author: Lorrie Miller Kohler

b. Do the same for those foods containing 5 or more milligrams of cholesterol. Record milligrams of cholesterol and quantity of food.

Cholesterol			

3. Select a high-fat food. Calculate % fat by weight in this food by dividing the weight of the fat (in grams) by the total weight of the food (in grams) and multiply by 100.

_____ grams/_____ grams =_____ x 100 =_____% fat by weight in _____ (selected food)

4. Calculate % calories from fat for this food.
    Step 1. Calculate the number of calories in the grams of fat this food contains.
    Step 2. Divide that number by total calories in the food.
    Step 3. Multiply by 100.

Calories from fat/total food cal. =_____ x 100% =_____ % cal. from fat in _____ (food)

5. Write a single sentence describing how the answer from Question #3 is different from the answer from Question #4.

6. What percent of total caloric intake for this day was contributed by this high-fat food?
    Calculate this by:
    Step 1. Multiplying grams of fat in food x 9 = calories from fat in the food.
    Step 2. Dividing this number by total number of calories for the day and multiplying by 100.

Calories from fat/total calories for day x 100 = _____% calories from fat in _____ (food)

7.   Note how Questions 3, 4, and 6 are asking different questions. As a group agree on and summarize the differences in a single sentence and write it below.

8.   What percent of the total calories come from saturated, monounsaturated, and polyunsaturated fat? The total grams of these fats is given in the Spreadsheet for Day 1. To calculate the amounts, multiply the number of grams of each type of fat by 9; then divide each answer by the total calories; and multiply by 100. Show your math.

_____% calories from saturated fat

_____% calories from monounsaturated fat

_____% calories from polyunsaturated fat

9.   How do the values in Question 8 compare with recommendations for total fat intake of 20%-35% of calories with less than 10% of calories coming from saturated fat? As a group summarize the comparisons in a sentence and write it here.

10.   Would you recommend any changes in the foods containing fats that were selected in this day's intake? If yes, what suggestion(s)? If not, why not? Record group's responses below.

Author: Lorrie Miller Kohler

11.　In Question #2 you listed foods containing cholesterol from in this day's intake. What is the origin—plant or animal—of these foods?

12.　Select 4 foods containing cholesterol and identify the specific source of the cholesterol in each.

Cholesterol-Containing Food	Source of Cholesterol

13.　How does the total cholesterol consumed in this day's intake compare with the recommendation to consume less than 300 milligrams of cholesterol a day?

14.　Would you make any suggestions for change in your cholesterol intake? If yes, what suggestion(s)? If no, why not? Record changes or reasons for no change below.

Author: Lorrie Miller Kohler　　　　110

Name_____

Diet Analysis Activity DA-8I: Protein Exercise

Objectives:
- To identify foods high in quantity of protein.
- To evaluate the adequacy of protein intake in grams and in the context of caloric intake.
- To compare the quality of protein in foods from plant and animal sources.
- To calculate percent calories from protein.
- To suggest menu choices that are high in protein and fiber and low in fat.

Instructions:
Answer the following questions about proteins using your 1-day Diet Analysis Plus printout.

1. Review the foods you ate using the Spreadsheet for Day 1 and identify the foods that contain more than 3 grams of protein. List these foods in decreasing order of amount of protein and according to their source. Also record quantity of food.

Animal-derived protein		Plant-derived protein	

Author: Lorrie Miller Kohler

2. Compare the quantity of protein obtained from these two sources. As a group agree on and describe in a single sentence the comparison. Record it below.

3. Describe and compare the total number of grams of protein (_____) with your RDA for protein (_____). * Remember the number of grams of protein consumed should be no more than twice the RDA for protein. What is twice the RDA for you?_____ Make this comparison in the context of looking at your calorie intake (_____) for the day and your RDA for calories (_____). In looking at the gram intake in the context of the caloric intake, is the quantity of your protein intake for this day adequate? Why or why not?

4. Each group **member** compare the quality of protein from the two sources (in Question #1). As a group describe the comparison in a single sentence. Record it below.

5.   In the space below <u>rank</u> all the protein-rich foods listed in Question #1 from <u>highest to lowest</u> according to the amount (in grams) of total fat. Record number of grams of fat.

6.   After looking at the data in Question #5, as a group, make a statement comparing the <u>source</u> of high protein-rich foods with their fat content.

7.   What percent of the total calories for the day came from protein? <u>Show your calculations below</u>.

Author: Lorrie Miller Kohler

8.   Do you think this day's intake implemented the Dietary Goals for protein? If yes, how? If no, how not? Record group members' responses.

9.   As a group delete one of this day's high-protein foods and add another protein source which **would not change the amount of protein, but would decrease the amount of total fat and increase the amount of fiber.** Use a food composition table to verify that your choice is a good one.

Name the food you are deleting. _____

*   How much protein does it contain? _____

*   How much fat does it contain? _____

*   How much fiber does it contain? _____

Name and give the amount of the food you are adding. _____

*   Indicate the amount of protein in the food you are adding. _____

*   Indicate the amount of fat in the food you are adding. _____

*   Indicate the amount of fiber in the food you are adding. _____

Name_____

DIET ANALYSIS ACTIVITY DA-9I: VITAMIN EXERCISE

Objectives:
- To identify food sources of selected vitamins.
- To use RDAs to evaluate intake of selected vitamins.
- To suggest menu choices that would provide adequate intake of selected vitamins.
- To identify major functions of selected vitamins.

Instructions:

Answer the following questions about vitamins using your 1-day Diet Analysis Plus printout.

1. Review the foods you ate using the Spreadsheet for Day 1 and identify the foods that contain 10 milligrams or more of vitamin C. List these foods in <u>decreasing order</u> of amount in milligrams of vitamin C they contain. Also record quantity of food.

2. Identify and list the foods that provide no vitamin C and determine their source (animal or plant). What do you observe about the source of vitamin- containing foods?

Author: Lorrie Miller Kohler

3. Compare and record the number of milligrams of vitamin C in your intake with the RDA. Is it too high, too low, or about right?

4. As a group delete one of this day's high vitamin C foods and add another that would not significantly change the amount of vitamin C but would <u>increase</u> the <u>amount of fiber</u>. Use a food composition table to verify that your choice is a good one.

Name the food you are deleting. _____

- Indicate the amount of vitamin C in deleted food. _____

- Indicate the amount of fiber in the deleted food. _____

Name and give the amount of food you are adding. _____

- Indicate the amount of Vitamin C in added food. _____

- Indicate the amount of fiber in the added food. _____

5. List three major functions of vitamin C in the body.

6. Identify the foods you ate that contain 10 or more REs of vitamin A. List them in order of decreasing amounts of RE of vitamin A they contain. Also record quantity of food. If the food contains preformed vitamin A place (**PA**) after the food, if it contains beta carotene, place (**BC**) after the food, and if it has been fortified with vitamin A, place (**F**) after the food.

7. Compare and record the number of REs of vitamin A in this day's intake with the RDA for vitamin A. Is it too high, too low, or about right?

8. Select a sweet food low in vitamin A, and identify a vitamin A-rich fruit that would be a good replacement.

9. Select a low-vitamin A vegetable and identify a vitamin A-rich vegetable that would be a good replacement.

Author: Lorrie Miller Kohler

10.   List three major functions of vitamin A in the body.

11.   For each of the following B vitamins, identify the two major sources in this day's diet.

thiamin (vit. B_1):

riboflavin (vit. B_2):

niacin (vit. B_3):

vitamin B_6:

vitamin B_{12}:

folate:

12.   As a group make a general statement about the major function of the B vitamins.

13.   Did you eat any processed foods high in vitamins? Why do you think such foods are high in many of the vitamins?

Name_____

DIET ANALYSIS ACTIVITY DA-10I: MINERAL EXERCISE

Objectives:
- To identify food sources of selected minerals.
- To use guidelines and recommendations to evaluate intake of selected minerals.
- To suggest menu choices that would provide adequate intake of selected minerals.
- To identify major functions of selected minerals.

Instructions:
Answer the following questions about minerals using your 1-day Diet Analysis Plus printout.

1. Review the foods you ate using the Spreadsheet for Day 1 and identify the foods that contain 10 milligrams (mg.) or more of calcium. List these foods in decreasing order of amount of calcium they contain. Also record quantity of food.

Foods containing 10 mgs. or more of calcium	

2. Compare and record the number of milligrams of calcium in this day's intake with your DRI (AI). Is it too high, too low, or about right?

3. As a group add a food of <u>plant</u> origin that is a good source of calcium. Use a food composition table to verify that your choice is a good one.

Name and give the amount of the food you are adding. _____

How much calcium is there in the food you are adding? _____

119   Author: Lorrie Miller Kohler

4.　　List two major functions of calcium in the body.

5.　　Identify the foods you ate that contain 0.7 milligrams (mg.) or more of iron. List them in order of decreasing amounts of milligrams of iron they contain. Also record quantity of food.

Foods containing 0.7 mgs. or more of iron		

6.　　Compare and record the number of milligrams of iron in this day's intake with your RDA. Is it too high, too low, or about right?

7.　　As a group change this day's intake to <u>increase</u> the amount of iron. List 3 additional food sources of iron and the amount of iron that each contains.

　　　1.

　　　2.

　　　3.

8.　　Describe the major function of iron in the body.

9.　　Are there any foods containing refined flour? Why do such foods have relatively high iron content?

10.　　What is the total amount of sodium consumed in this day?

Author: Lorrie Miller Kohler　　　　　120

11.   Do you think that this intake is high, low, or about right? Upon what guideline or recommendation (other than the computer printout) are you basing your opinion?

12.   Carefully review the foods and their sodium content. As a group suggest <u>three</u> specific menu changes that would lower the sodium content of this day's intake. List the <u>food</u> and the <u>amount</u> of sodium it contains.

    1.

    2.

    3.

        Author: Lorrie Miller Kohler

COOPERATIVE GROUP'S EVALUATION FORM

At the completion of your Base Group's activity, complete this evaluation form. The Reader/ Recorder reads the questions to the group, and records the group's agreed upon answers. Place the Evaluation Form into the group's folder and turn it in with the completed group activity.

1. Overall, <u>how effectively</u> did your group work together on this assignment? (Circle the appropriate response.)

1	2	3	4	5
not at all	poorly	adequately	well	extremely well

2. <u>How many</u> of the total number of group members <u>participated actively</u> most of the time? (Circle the appropriate number.)

    0   1   2   3   4

3. <u>How many</u> of the group's members <u>were fully prepared</u> for the group work most of the time? (Circle the appropriate number.)

    0   1   2   3   4

4. Each person give one specific example of <u>something you learned from the group</u> that you probably wouldn't have learned on your own.

5. Suggest <u>one specific practical change</u> the group could make that would help improve everyone's learning.

6. Describe how well each group member carried out his or her respective role.

Exercises #1, #2, #3, #6 and #7 are designed for use with pre-set menus. The Diet Analysis Plus reports for these exercises are included in this workbook. Exercises #4, #5 and #8 are to be used with the student's own Diet Analysis Plus reports for 3 days plus the average of these 3 days, and guide the student through an investigation of his or her own intake and its nutritional implications.

Author: Elaine M. Long

Use the menu below to answer the following questions.

4 pancakes (4" in diameter, 1 1/3 oz. per pancake)
2 tsp. margarine (soft)
6 tsp. syrup
1 cup orange juice
1 cup skim milk

2 soft shell tacos, containing:
  2 flour tortillas (8 inch across, 1 ¾ oz./tortilla)
  2 oz. lean ground beef
  2 oz. mozzarella cheese
  1 cup shredded romaine lettuce
  1 sliced tomato (medium size)
  1 tbsp. sliced onion
  3 tbsp. taco sauce

2 cups reduced fat (2%) milk
6 small Vanilla Wafers

3 oz. broiled chicken breast (w/o skin)
1 large baked potato (weight 7 oz.)
2 tbsp. sour cream (fat-free)
1 cup peas (peas are "starchy" vegetables)

1 large bran muffin (weight 4 oz.)
1 cup apple juice

1.   Identify the main sources of carbohydrate in this menu.

2.   Divide your carbohydrate sources into foods containing mostly complex carbohydrates or mostly simple carbohydrates.

3.   In your list of simple carbohydrates, which ones would you classify as naturally occurring sugars and which ones would you classify as refined or added sugars? Why?

4.   In your list of complex carbohydrates, which ones would you classify as significant sources of dietary fiber? Why?

                    Author: Elaine M. Long

5.   Next estimate the total grams of carbohydrate in this menu. To do this you need to determine: (a) the exchange list and (b) corresponding number of exchanges for each item.

6.   Using the exchange lists, estimate the total grams of carbohydrate in this menu.

7.   Referring to the food sources of carbohydrate in this menu, what are two strengths of this menu? Answer this question in terms of the dietary guidelines and the diet planning principles.

8.   What suggestions do you have to improve the food sources of carbohydrates (not the fats) in this menu? Be specific.

Student Name: Exercise #1 - Carbohydrates
Student ID #:
Instructor Name:
Class Days:
Class Time:

| | | | | Spreadsheet | | | |
Amount	Food Item	Weight (g)	Cals	H2O (g)	Prot (g)	Carb (g)	Fiber (g)
4 each	Plain Pancakes-Mix-Prepared 4"	152.00	294.88	80.56	7.90	55.78	1.98
2 tsp	MargarineSpread-Hy(Soy+Cttnsd)60%Fat	9.60	51.84	3.55	0.06	0	0
6 tsp	Pancake Syrup	40.00	114.80	9.64	0	30.28	0
1 cup	Fresh Orange Juice	248.00	111.60	218.98	1.74	25.79	0.50
1 cup	Nonfat SkimMilk-Protein Fort.W/Add Vit A	246.00	100.86	219.83	9.74	13.68	0
2 each	Mission Flour Tortillas- Soft Taco- 8"	102.00	292.74	34.27	8.87	50.59	--
2 oz-wt	Ground Beef-Extra Lean (9% Fat) Raw	56.70	95.82	39.35	11.62	0	0
2 oz-wt	Mozzarella Cheese-Whole Milk-Shredded	56.70	159.33	30.70	11.01	1.26	0
1 cup	Romaine (Cos) Lettuce-Chopped-Cup	56.00	7.84	53.15	0.91	1.33	0.95
1 cup	Tomatoes-Chopped/Sliced,Red,Raw,Ripe-Cup	180.00	37.80	168.77	1.53	8.35	1.98
1 tbs	Yellow Onions-Raw Slices-Cup	7.19	2.73	6.45	0.08	0.62	0.13
3 tbs	Ortega Taco Sauce-Hot-Thick&Smooth NFC	48.00	30.00	40.83	0	6.00	0
2 cup	2% Fat Milk-Vitamin A Added	488.00	244.00	435.34	16.25	23.42	0
6 each	Nabisco (Vanilla) Nilla Wafers Cookies	24.00	105.00	--	0.75	18.00	0.38
3 oz-wt	Chicken Breast-w/o Skin-Boneless-Roasted	85.05	140.33	55.50	26.38	0	0
7 oz-wt	Baked Potatow/Skin+Flesh-Medium	198.45	184.56	148.62	4.96	41.97	4.37
2 tbs	Nonfat Sour Cream DGI	32.00	25.00	25.38	2.00	4.00	0
1 cup	Bird's Eye Green Peas DFV	178.00	141.65	140.44	9.74	25.24	9.43
4 oz-wt	Bran Muffin-Prep f/Recipe w/Buttermilk	113.40	319.51	39.80	7.51	47.12	6.54
1 cup	Apple Juice-Canned/Bottled,Unsweetened	248.00	116.56	218.07	0.15	28.97	0.25
	Totals	2569.09	2576.85	1969.23	121.20	382.40	26.50

Student Name: Exercise #1 - Carbohydrates
Student ID #:
Instructor Name:
Class Days:
Class Time:

Spreadsheet

Amount	Food Item	Fat (g)	Sat (g)	Mono (g)	Poly (g)	Chol (mg)	A-RE (RE)
4 each	Plain Pancakes-Mix-Prepared 4"	3.80	0.77	1.34	1.25	18.24	15.20
2 tsp	MargarineSpread-Hy(Soy+Cttnsd)60%Fat	5.84	1.15	3.73	0.68	0	76.80
6 tsp	Pancake Syrup	0	0	0	0	0	0
1 cup	Fresh Orange Juice	0.50	0.06	0.09	0.10	0	49.60
1 cup	Nonfat SkimMilk-Protein Fort.W/Add Vit A	0.61	0.40	0.16	0.02	4.92	150.06
2 each	Mission Flour Tortillas- Soft Taco- 8"	6.12	0.71	2.83	0.94	--	--
2 oz-wt	Ground Beef-Extra Lean (9% Fat) Raw	5.11	2.04	2.22	0.21	20.64	0
2 oz-wt	Mozzarella Cheese-Whole Milk-Shredded	12.25	7.46	3.73	0.43	44.23	120.77
1 cup	Romaine (Cos) Lettuce-Chopped-Cup	0.11	0.01	0.00	0.06	0	145.60
1 cup	Tomatoes-Chopped/Sliced,Red,Raw,Ripe-Cup	0.59	0.08	0.09	0.24	0	111.60
1 tbs	Yellow Onions-Raw Slices-Cup	0.01	0.00	0.00	0.00	0	0
3 tbs	Ortega Taco Sauce-Hot-Thick&Smooth NFC	0	0	0	0	0	0
2 cup	2% Fat Milk-Vitamin A Added	9.37	5.83	2.71	0.35	39.04	278.16
6 each	Nabisco (Vanilla) Nilla Wafers Cookies	3.75	0.75	1.13	0	1.88	--
3 oz-wt	Chicken Breast-w/o Skin-Boneless-Roasted	3.04	0.86	1.05	0.65	72.29	5.10
7 oz-wt	Baked Potatow/Skin+Flesh-Medium	0.26	0.07	0.01	0.12	0	3.97
2 tbs	Nonfat Sour Cream DGI	0	0	0	0	5.00	40.00
1 cup	Bird's Eye Green Peas DFV	0.91	0.16	--	--	0	124.39
4 oz-wt	Bran Muffin-Prep f/Recipe w/Buttermilk	13.31	1.33	--	--	35.21	10.01
1 cup	Apple Juice-Canned/Bottled,Unsweetened	0.27	0.05	0.01	0.08	0	0.25
	Totals	65.84	21.73	19.10	5.14	241.44	1131.51

Student Name: Exercise #1 - Carbohydrates
Student ID #:
Instructor Name:
Class Days:
Class Time:

Spreadsheet

Amount	Food Item	B1 (mg)	B2 (mg)	B3 (mg)	B6 (mg)	B12 (mcg)	Fola (mcg)
4 each	Plain Pancakes-Mix-Prepared 4"	0.32	0.33	2.60	0.14	0.30	56.24
2 tsp	MargarineSpread-Hy(Soy+Cttnsd)60%Fat	0.00	0.00	0.00	0.00	0.01	0.10
6 tsp	Pancake Syrup	0.00	0.00	0.01	0	0	0
1 cup	Fresh Orange Juice	0.22	0.07	0.99	0.10	0	74.40
1 cup	Nonfat SkimMilk-Protein Fort.W/Add Vit A	0.11	0.48	0.25	0.12	1.06	14.76
2 each	Mission Flour Tortillas- Soft Taco- 8"	--	--	--	--	--	--
2 oz-wt	Ground Beef-Extra Lean (9% Fat) Raw	0.04	0.16	2.82	0.16	1.29	4.98
2 oz-wt	Mozzarella Cheese-Whole Milk-Shredded	0.01	0.14	0.05	0.03	0.37	3.97
1 cup	Romaine (Cos) Lettuce-Chopped-Cup	0.06	0.06	0.28	0.03	0	76.16
1 cup	Tomatoes-Chopped/Sliced,Red,Raw,Ripe-Cup	0.11	0.09	1.13	0.14	0	27.00
1 tbs	Yellow Onions-Raw Slices-Cup	0.00	0.00	0.01	0.01	0	1.37
3 tbs	Ortega Taco Sauce-Hot-Thick&Smooth NFC	--	--	--	--	--	--
2 cup	2% Fat Milk-Vitamin A Added	0.19	0.81	0.42	0.21	1.76	24.40
6 each	Nabisco (Vanilla) Nilla Wafers Cookies	--	--	--	--	--	--
3 oz-wt	Chicken Breast-w/o Skin-Boneless-Roasted	0.06	0.10	11.66	0.51	0.29	3.40
7 oz-wt	Baked Potatow/Skin+Flesh-Medium	0.13	0.10	2.80	0.62	0	55.57
2 tbs	Nonfat Sour Cream DGI	--	--	--	--	--	--
1 cup	Bird's Eye Green Peas DFV	0.55	0.18	3.92	0.23	0	98.43
4 oz-wt	Bran Muffin-Prep f/Recipe w/Buttermilk	0.45	0.24	2.00	0.05	0.08	33.92
1 cup	Apple Juice-Canned/Bottled,Unsweetened	0.05	0.04	0.25	0.07	0	0
	Totals	2.30	2.78	29.18	2.43	5.15	474.69

Student Name: Exercise #1 - Carbohydrates
Student ID #:
Instructor Name:
Class Days:
Class Time:

Spreadsheet

Amount	Food Item	Vit C (mg)	D-mcg (mcg)	E-aTE (mg)	Calc (mg)	Iron (mg)	Mag (mg)
4 each	Plain Pancakes-Mix-Prepared 4"	0.30	1.93	1.29	191.52	2.37	30.40
2 tsp	MargarineSpread-Hy(Soy+Cttnsd)60%Fat	0.01	--	0.48	2.02	0	0.19
6 tsp	Pancake Syrup	0	0	0	0.40	0.04	0.80
1 cup	Fresh Orange Juice	124.00	0	0.22	27.28	0.50	27.28
1 cup	Nonfat SkimMilk-Protein Fort.W/Add Vit A	2.71	2.46	0.10	351.78	0.15	39.36
2 each	Mission Flour Tortillas- Soft Taco- 8"	--	--	--	194.82	2.03	
2 oz-wt	Ground Beef-Extra Lean (9% Fat) Raw	0	0.19	0.10	4.35	1.22	12.47
2 oz-wt	Mozzarella Cheese-Whole Milk-Shredded	0	0.09	0.20	293.14	0.10	10.77
1 cup	Romaine (Cos) Lettuce-Chopped-Cup	13.44	0	0.25	20.16	0.62	3.36
1 cup	Tomatoes-Chopped/Sliced,Red,Raw,Ripe-Cup	34.38	0	0.68	9.00	0.81	19.80
1 tbs	Yellow Onions-Raw Slices-Cup	0.46	0	0.02	1.44	0.02	0.72
3 tbs	Ortega Taco Sauce-Hot-Thick&Smooth NFC	0	--	--	0	1.08	--
2 cup	2% Fat Milk-Vitamin A Added	4.88	4.88	0.34	595.36	0.24	68.32
6 each	Nabisco (Vanilla) Nilla Wafers Cookies	--	--	--	15.00	0.81	--
3 oz-wt	Chicken Breast-w/o Skin-Boneless-Roasted	0	0.26	0.23	12.76	0.88	24.66
7 oz-wt	Baked Potatow/Skin+Flesh-Medium	19.05	0	0.08	29.77	2.14	55.57
2 tbs	Nonfat Sour Cream DGI	0	0	--	60.00	0	--
1 cup	Bird's Eye Green Peas DFV	35.51	--	--	42.35	2.83	44.50
4 oz-wt	Bran Muffin-Prep f/Recipe w/Buttermilk	0.37	0.10	2.57	108.74	2.61	13.33
1 cup	Apple Juice-Canned/Bottled,Unsweetened	2.23	0	0.02	17.36	0.92	7.44
	Totals	237.34	9.90	6.58	1977.24	19.37	358.98

Student Name: Exercise #1 - Carbohydrates
Student ID #:
Instructor Name:
Class Days:
Class Time:

| | | | | | | Spreadsheet | |
Amount	Food Item	Phos (mg)	Potas (mg)	Sod (mg)	Zinc (mg)	Caff (mg)	Alco (g)
4 each	Plain Pancakes-Mix-Prepared 4"	507.68	266.00	954.56	0.59	0	0
2 tsp	MargarineSpread-Hy(Soy+Cttnsd)60%Fat	1.54	2.88	95.42	0	0	0
6 tsp	Pancake Syrup	3.60	0.80	33.20	0.02	0	0
1 cup	Fresh Orange Juice	42.16	496.00	2.48	0.12	0	0
1 cup	Nonfat SkimMilk-Protein Fort.W/Add Vit A	275.52	447.72	145.14	1.11	0	0
2 each	Mission Flour Tortillas- Soft Taco- 8"	--	--	497.76	--	0	0
2 oz-wt	Ground Beef-Extra Lean (9% Fat) Raw	87.89	176.90	41.05	2.59	0	0
2 oz-wt	Mozzarella Cheese-Whole Milk-Shredded	210.36	37.99	211.49	1.25	0	0
1 cup	Romaine (Cos) Lettuce-Chopped-Cup	25.20	162.40	4.48	0.14	0	0
1 cup	Tomatoes-Chopped/Sliced,Red,Raw,Ripe-Cup	43.20	399.60	16.20	0.16	0	0
1 tbs	Yellow Onions-Raw Slices-Cup	2.37	11.28	0.22	0.01	0	0
3 tbs	Ortega Taco Sauce-Hot-Thick&Smooth NFC	--	120.00	360.00	--	0	0
2 cup	2% Fat Milk-Vitamin A Added	463.60	751.52	244.00	1.90	0	0
6 each	Nabisco (Vanilla) Nilla Wafers Cookies	--	22.50	75.00	--	0	0
3 oz-wt	Chicken Breast-w/o Skin-Boneless-Roasted	193.91	217.73	62.94	0.85	0	0
7 oz-wt	Baked Potatow/Skin+Flesh-Medium	138.92	1061.71	19.85	0.71	0	0
2 tbs	Nonfat Sour Cream DGI	--	--	50.00	--	0	0
1 cup	Bird's Eye Green Peas DFV	158.42	304.38	249.02	1.78	0	0
4 oz-wt	Bran Muffin-Prep f/Recipe w/Buttermilk	64.48	396.09	507.82	0.28	0	0
1 cup	Apple Juice-Canned/Bottled,Unsweetened	17.36	295.12	7.44	0.07	0	0
	Totals	2236.20	5170.62	3578.07	11.60	0	0

Student Name: Exercise #1 - Carbohydrates
Student ID #:
Instructor Name:
Class Days:
Class Time:

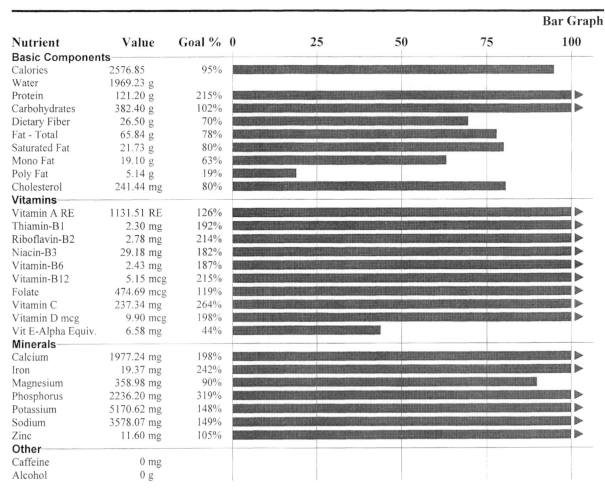

Nutrient	Value	Goal %
Basic Components		
Calories	2576.85	95%
Water	1969.23 g	
Protein	121.20 g	215%
Carbohydrates	382.40 g	102%
Dietary Fiber	26.50 g	70%
Fat - Total	65.84 g	78%
Saturated Fat	21.73 g	80%
Mono Fat	19.10 g	63%
Poly Fat	5.14 g	19%
Cholesterol	241.44 mg	80%
Vitamins		
Vitamin A RE	1131.51 RE	126%
Thiamin-B1	2.30 mg	192%
Riboflavin-B2	2.78 mg	214%
Niacin-B3	29.18 mg	182%
Vitamin-B6	2.43 mg	187%
Vitamin-B12	5.15 mcg	215%
Folate	474.69 mcg	119%
Vitamin C	237.34 mg	264%
Vitamin D mcg	9.90 mcg	198%
Vit E-Alpha Equiv.	6.58 mg	44%
Minerals		
Calcium	1977.24 mg	198%
Iron	19.37 mg	242%
Magnesium	358.98 mg	90%
Phosphorus	2236.20 mg	319%
Potassium	5170.62 mg	148%
Sodium	3578.07 mg	149%
Zinc	11.60 mg	105%
Other		
Caffeine	0 mg	
Alcohol	0 g	

Exercise #2: Dietary Fats and Cholesterol

Use the menu below to answer the following questions.

1 ½ cups Kix cereal (weight ~1 oz.)
1 small banana (5 oz.)
1 cup reduced-fat (2%) milk
1 cup coffee
2 tsp. half and half
1 slice whole-wheat toast
1 tsp. margarine (all vegetable, not diet)

Bacon/lettuce/tomato sandwich, containing:
   4 slices bacon (20 slices/lb.)
   1 tbsp. mayonnaise
   ¼ cup romaine lettuce
   2 slices tomato (¼-inch thick)
   2 slices toasted rye bread

Salad, containing:
   1 cup romaine lettuce (chopped)
   1 oz. grated cheddar cheese
   1 tbsp. grated carrots
   1 tbsp. cucumber
   2 tbsp sunflower seeds
   ¼ of a medium avocado
   ½ chopped egg
   1 slice bacon (crumbled)
   3 tbsp. Ranch dressing (regular, not low fat)

3 cups ice tea with lemon

16 wheat thins
2 tbsp. creamy peanut butter

Stir-fry, containing:
   3-ounce round steak
   1 cup cooked brown rice
   ¼ cup broccoli
   ¼ cup carrots
   ¼ cup chopped green pepper
   ¼ cup chopped mushrooms

1 cup reduced-fat (2%) milk

1 cup vanilla ice cream

1.   Identity the main sources of fat in this menu.

2.   Classify your fat sources as sources of either invisible of visible fat. Foods classified as visible fats generally have 90% or more of their kcal from fat. Many sources of invisible fat contain 50 - 80% of their kcal from fat.

3.   Classify your fat sources as containing predominately saturated fatty acids, monounsaturated fatty acids or polyunsaturated fatty acids.

Author: Elaine M. Long

4. What fat sources in this menu are high in the essential fatty acids, linoleic and linolenic? Note these are also referred to as omega-6 and omega-3 fatty acids. What other foods in this menu supply omega-3 fatty acids?

5. Identify the food sources of cholesterol in this menu.

6. Are there any sources of trans fatty acids in this menu? If yes, list them.

7. Using the Exchange lists, estimate the total grams of fat in this menu.

8. In terms of the diet planning principles and dietary guidelines, what are some of the strengths of this menu?

9. Based on the detailed information about the sources and types of lipids in this menu, what specific changes would you suggest?

Student Name: Exercise#2 - Dietary Fats & Cholesterol
Student ID #:
Instructor Name:
Class Days:
Class Time:

							Spreadsheet
		Weight	Cals	H2O	Prot	Carb	Fiber
Amount	Food Item	(g)		(g)	(g)	(g)	(g)
1.5 cup	General Mills Kix Cereal GML	33.83	127.56	0.73	2.03	29.10	1.02
5 oz-wt	Banana Slices-Cup	141.75	130.41	105.26	1.46	33.21	3.40
1 cup	2% Fat Milk-Vitamin A Added	244.00	122.00	217.67	8.13	11.71	0
1 cup	Brewed Coffee-Prep w/Tap Water	237.00	4.74	235.34	0.24	0.95	0
2 tsp	Half & Half Cream	10.00	13.00	8.06	0.30	0.43	0
1 piece	Whole Wheat Bread-Slice-Toasted	25.00	69.25	7.50	2.73	12.93	1.85
1 tsp	Parkay Margarine-70% Veg Oil-Stick	4.67	30.00	--	0	0	0
3.2 oz-wt	Bacon-Svg JBX	90.72	518.40	--	51.84	0	0
1 tbs	Mayonnaise-(Soybean Oil) w/Salt	13.80	98.95	2.11	0.15	0.37	0
0.25 cup	Romaine (Cos) Lettuce-Chopped-Cup	14.00	1.96	13.29	0.23	0.33	0.24
2 each	Tomatoes-LrgSlice-Sandwich Component SUB	29.00	6.00	27.88	0	1.00	0
2 piece	Rye Bread-Toasted-Slice-Reg	48.00	136.32	14.88	4.51	25.49	3.07
1 cup	Romaine (Cos) Lettuce-Chopped-Cup	56.00	7.84	53.15	0.91	1.33	0.95
1 oz-wt	Cheddar Cheese-Shredded	28.35	114.25	10.42	7.06	0.36	0
1 tbs	Carrots-Raw-Grated-Cup	6.88	2.96	6.04	0.07	0.70	0.21
0.5 oz-wt	Cucumber-Raw, Medium FDA	14.18	2.15	13.54	0.14	0.43	0.14
2 tbs	Sunflower Seeds FRL	18.67	120.00	0.19	4.67	3.33	1.33
0.25 each	California Avocado-Each	43.25	76.55	31.38	0.91	2.99	2.12
0.5 each	Large Poached Egg-Each	25.00	37.25	18.76	3.11	0.30	0
0.8 oz-wt	Bacon-Svg JBX	22.68	129.60	--	12.96	0	0
3 tbs	Marie's Buttermilk RanchDressing/Dip DFV	45.00	270.00	9.75	0	6.00	0
3 cup	Brewed Tea	710.40	7.10	708.27	0	2.13	0
0.5 tsp	Fresh Lemon Juice	2.54	0.64	2.31	0.01	0.22	0.01
16 each	Wheat Thins Crackers-Low Sodium	32.00	146.88	1.02	2.69	21.66	0.64
2 tbs	Creamy Peanut Butter	32.00	189.76	0.39	8.07	6.17	1.89
3 oz-wt	Beef Roast-BottomRound-All-1/4"Trim-Rstd	85.05	233.89	44.35	24.38	0	0
1 cup	Long Grain Brown Rice-Ckd	195.00	216.45	142.53	5.03	44.77	3.51
0.25 cup	Broccoli Pieces-Stir Fried	39.00	10.92	35.37	1.17	2.05	1.17
0.25 cup	Carrot Slices-Stir Fried	39.00	16.81	34.20	0.41	3.94	1.17
2 oz-wt	Med Sweet Green Bell Peppers-Raw-Each	56.70	15.31	52.27	0.50	3.65	1.02
0.25 cup	Mushroom Pieces/Slices-Raw-Cup	17.50	4.38	16.07	0.51	0.71	0.21
1 cup	2% Fat Milk-Vitamin A Added	244.00	122.00	217.67	8.13	11.71	0
1 cup	Vanilla Ice Cream	132.00	265.32	80.52	4.62	31.15	0.92
	Totals	2736.96	3248.63	2110.92	156.93	259.13	24.87

Student Name:　　Exercise#2 - Dietary Fats & Cholesterol
Student ID #:
Instructor Name:
Class Days:
Class Time:

Spreadsheet

Amount	Food Item	Fat (g)	Sat (g)	Mono (g)	Poly (g)	Chol (mg)	A-RE (RE)
1.5 cup	General Mills Kix Cereal GML	0.68	0.17	0.18	0.23	0	179.32
5 oz-wt	Banana Slices-Cup	0.68	0.26	0.06	0.13	0	11.34
1 cup	2% Fat Milk-Vitamin A Added	4.68	2.92	1.35	0.17	19.52	139.08
1 cup	Brewed Coffee-Prep w/Tap Water	0.01	0.00	0	0.00	0	0
2 tsp	Half & Half Cream	1.15	0.72	0.33	0.04	3.70	10.70
1 piece	Whole Wheat Bread-Slice-Toasted	1.20	0.26	0.47	0.28	0	0
1 tsp	Parkay Margarine-70% Veg Oil-Stick	3.33	0.67	--	--	0	33.33
3.2 oz-wt	Bacon-Svg JBX	38.88	12.96	--	--	129.60	0
1 tbs	Mayonnaise-(Soybean Oil) w/Salt	10.96	1.63	3.13	5.70	8.14	11.59
0.25 cup	Romaine (Cos) Lettuce-Chopped-Cup	0.03	0.00	0.00	0.01	0	36.40
2 each	Tomatoes-LrgSlice-Sandwich Component SUB	0	0	0	0	0	32.80
2 piece	Rye Bread-Toasted-Slice-Reg	1.73	0.33	0.69	0.42	0	--
1 cup	Romaine (Cos) Lettuce-Chopped-Cup	0.11	0.01	0.00	0.06	0	145.60
1 oz-wt	Cheddar Cheese-Shredded	9.40	5.98	2.66	0.27	29.77	78.81
1 tbs	Carrots-Raw-Grated-Cup	0.01	0.00	0.00	0.01	0	193.33
0.5 oz-wt	Cucumber-Raw, Medium FDA	0	0	0	0	0	2.86
2 tbs	Sunflower Seeds FRL	10.00	1.00	--	--	0	0
0.25 each	California Avocado-Each	7.50	1.12	4.85	0.88	0	26.82
0.5 each	Large Poached Egg-Each	2.50	0.77	0.95	0.34	105.75	47.50
0.8 oz-wt	Bacon-Svg JBX	9.72	3.24	--	--	32.40	0
3 tbs	Marie's Buttermilk RanchDressing/Dip DFV	27.00	4.50	--	--	22.50	0
3 cup	Brewed Tea	0.04	0.01	0.01	0.03	0	0
0.5 tsp	Fresh Lemon Juice	0	0	0	0	0	0.05
16 each	Wheat Thins Crackers-Low Sodium	5.92	2.00	2.33	1.26	0	4.80
2 tbs	Creamy Peanut Butter	16.33	3.31	7.77	4.41	0	0
3 oz-wt	Beef Roast-BottomRound-All-1/4"Trim-Rstd	14.37	5.42	6.25	0.54	81.65	0
1 cup	Long Grain Brown Rice-Ckd	1.75	0.35	0.64	0.63	0	0
0.25 cup	Broccoli Pieces-Stir Fried	0.14	0.02	0.01	0.07	0	54.02
0.25 cup	Carrot Slices-Stir Fried	0.07	0.01	0.00	0.03	0	988.65
2 oz-wt	Med Sweet Green Bell Peppers-Raw-Each	0.11	0.02	0.01	0.06	0	36.29
0.25 cup	Mushroom Pieces/Slices-Raw-Cup	0.06	0.01	0.00	0.02	0	0
1 cup	2% Fat Milk-Vitamin A Added	4.68	2.92	1.35	0.17	19.52	139.08
1 cup	Vanilla Ice Cream	14.52	8.96	4.18	0.54	58.08	159.72
	Totals	187.56	59.57	37.24	16.32	510.63	2332.09

Student Name: Exercise#2 - Dietary Fats & Cholesterol
Student ID #:
Instructor Name:
Class Days:
Class Time:

Spreadsheet

Amount	Food Item	B1 (mg)	B2 (mg)	B3 (mg)	B6 (mg)	B12 (mcg)	Fola (mcg)
1.5 cup	General Mills Kix Cereal GML	0.42	0.48	5.65	0.57	1.69	225.68
5 oz-wt	Banana Slices-Cup	0.06	0.14	0.77	0.82	0	26.93
1 cup	2% Fat Milk-Vitamin A Added	0.10	0.40	0.21	0.10	0.88	12.20
1 cup	Brewed Coffee-Prep w/Tap Water	0	0	0.53	0	0	0
2 tsp	Half & Half Cream	0.00	0.01	0.01	0.00	0.03	0.30
1 piece	Whole Wheat Bread-Slice-Toasted	0.08	0.05	0.97	0.05	0.00	9.75
1 tsp	Parkay Margarine-70% Veg Oil-Stick	--	--	--	--	--	--
3.2 oz-wt	Bacon-Svg JBX	--	--	--	--	--	--
1 tbs	Mayonnaise-(Soybean Oil) w/Salt	0	0	0.00	0.08	0.04	1.10
0.25 cup	Romaine (Cos) Lettuce-Chopped-Cup	0.01	0.01	0.07	0.01	0	19.04
2 each	Tomatoes-LrgSlice-Sandwich Component SUB	--	--	--	--	0	--
2 piece	Rye Bread-Toasted-Slice-Reg	0.18	0.16	1.81	0.04	0	35.04
1 cup	Romaine (Cos) Lettuce-Chopped-Cup	0.06	0.06	0.28	0.03	0	76.16
1 oz-wt	Cheddar Cheese-Shredded	0.01	0.11	0.02	0.02	0.24	5.10
1 tbs	Carrots-Raw-Grated-Cup	0.01	0.00	0.06	0.01	0	0.96
0.5 oz-wt	Cucumber-Raw, Medium FDA	--	--	--	--	0	--
2 tbs	Sunflower Seeds FRL	--	--	--	--	0	--
0.25 each	California Avocado-Each	0.05	0.05	0.83	0.12	0	28.54
0.5 each	Large Poached Egg-Each	0.01	0.11	0.02	0.03	0.20	8.75
0.8 oz-wt	Bacon-Svg JBX	--	--	--	--	--	--
3 tbs	Marie's Buttermilk RanchDressing/Dip DFV	--	--	--	--	--	--
3 cup	Brewed Tea	0	0.10	0	0	0	35.52
0.5 tsp	Fresh Lemon Juice	0.00	0.00	0.00	0.00	0	0.33
16 each	Wheat Thins Crackers-Low Sodium	0.15	0.11	1.51	0.04	0	8.32
2 tbs	Creamy Peanut Butter	0.03	0.03	4.29	0.15	0	23.68
3 oz-wt	Beef Roast-BottomRound-All-1/4"Trim-Rstd	0.06	0.20	3.17	0.28	2.00	8.51
1 cup	Long Grain Brown Rice-Ckd	0.19	0.05	2.98	0.28	0	7.80
0.25 cup	Broccoli Pieces-Stir Fried	0.02	0.04	0.24	0.06	0	22.11
0.25 cup	Carrot Slices-Stir Fried	0.03	0.02	0.34	0.05	0	5.19
2 oz-wt	Med Sweet Green Bell Peppers-Raw-Each	0.04	0.02	0.29	0.14	0	12.47
0.25 cup	Mushroom Pieces/Slices-Raw-Cup	0.02	0.07	0.71	0.02	0.01	2.10
1 cup	2% Fat Milk-Vitamin A Added	0.10	0.40	0.21	0.10	0.88	12.20
1 cup	Vanilla Ice Cream	0.05	0.32	0.15	0.06	0.51	5.28
	Totals	1.68	2.96	25.11	3.06	6.48	593.07

Student Name: Exercise#2 - Dietary Fats & Cholesterol
Student ID #:
Instructor Name:
Class Days:
Class Time:

| | | | | | | | **Spreadsheet** |
Amount	Food Item	Vit C (mg)	D-mcg (mcg)	E-aTE (mg)	Calc (mg)	Iron (mg)	Mag (mg)
1.5 cup	General Mills Kix Cereal GML	7.11	1.18	0.09	169.17	9.14	9.14
5 oz-wt	Banana Slices-Cup	12.90	0	0.38	8.50	0.44	41.11
1 cup	2% Fat Milk-Vitamin A Added	2.44	2.44	0.17	297.68	0.12	34.16
1 cup	Brewed Coffee-Prep w/Tap Water	0	0	0	4.74	0.12	11.85
2 tsp	Half & Half Cream	0.09	0.04	0.01	10.50	0.01	1.00
1 piece	Whole Wheat Bread-Slice-Toasted	0	0.05	0.29	20.25	0.93	24.25
1 tsp	Parkay Margarine-70% Veg Oil-Stick	0	--	--	0	0	--
3.2 oz-wt	Bacon-Svg JBX	0	--	--	0	0	--
1 tbs	Mayonnaise-(Soybean Oil) w/Salt	0	0.04	0.32	2.48	0.07	0.14
0.25 cup	Romaine (Cos) Lettuce-Chopped-Cup	3.36	0	0.06	5.04	0.15	0.84
2 each	Tomatoes-LrgSlice-Sandwich Component SUB	5.00	--	--	2.00	0	--
2 piece	Rye Bread-Toasted-Slice-Reg	0.10	0	0.29	38.40	1.49	20.64
1 cup	Romaine (Cos) Lettuce-Chopped-Cup	13.44	0	0.25	20.16	0.62	3.36
1 oz-wt	Cheddar Cheese-Shredded	0	0.09	0.10	204.40	0.19	7.94
1 tbs	Carrots-Raw-Grated-Cup	0.64	0	0.03	1.86	0.03	1.03
0.5 oz-wt	Cucumber-Raw, Medium FDA	0.86	--	--	2.86	0.05	--
2 tbs	Sunflower Seeds FRL	0	--	--	13.33	0.96	--
0.25 each	California Avocado-Each	3.42	0	0.58	4.76	0.51	17.73
0.5 each	Large Poached Egg-Each	0	0.33	0.26	12.25	0.36	2.50
0.8 oz-wt	Bacon-Svg JBX	0	--	--	0	0	--
3 tbs	Marie's Buttermilk RanchDressing/Dip DFV	0	--	--	0	0	--
3 cup	Brewed Tea	0	0	0	0	0.14	21.31
0.5 tsp	Fresh Lemon Juice	1.17	0	0.00	0.18	0.00	0.15
16 each	Wheat Thins Crackers-Low Sodium	0	0	0.19	10.24	1.27	16.64
2 tbs	Creamy Peanut Butter	0	0	3.20	12.16	0.59	50.88
3 oz-wt	Beef Roast-BottomRound-All-1/4"Trim-Rstd	0	0.26	0.16	5.10	2.65	18.71
1 cup	Long Grain Brown Rice-Ckd	0	0	0.42	19.50	0.82	83.85
0.25 cup	Broccoli Pieces-Stir Fried	30.85	0	0.19	18.68	0.34	9.75
0.25 cup	Carrot Slices-Stir Fried	2.91	0	0.16	10.53	0.20	5.85
2 oz-wt	Med Sweet Green Bell Peppers-Raw-Each	50.63	0	0.39	5.10	0.26	5.67
0.25 cup	Mushroom Pieces/Slices-Raw-Cup	0.40	0.33	0.02	0.88	0.18	1.75
1 cup	2% Fat Milk-Vitamin A Added	2.44	2.44	0.17	297.68	0.12	34.16
1 cup	Vanilla Ice Cream	0.79	0.15	0.28	168.96	0.12	18.48
	Totals	138.54	7.35	8.02	1367.41	21.88	442.89

Student Name: Exercise#2 - Dietary Fats & Cholesterol
Student ID #:
Instructor Name:
Class Days:
Class Time:

		Spreadsheet					
Amount	Food Item	Phos (mg)	Potas (mg)	Sod (mg)	Zinc (mg)	Caff (mg)	Alco (g)
1.5 cup	General Mills Kix Cereal GML	45.00	39.59	301.47	4.23	0	0
5 oz-wt	Banana Slices-Cup	28.35	561.33	1.42	0.23	0	0
1 cup	2% Fat Milk-Vitamin A Added	231.80	375.76	122.00	0.95	0	0
1 cup	Brewed Coffee-Prep w/Tap Water	2.37	127.98	4.74	0.05	137.46	0
2 tsp	Half & Half Cream	9.50	13.00	4.10	0.05	0	0
1 piece	Whole Wheat Bread-Slice-Toasted	64.50	70.75	148.00	0.55	0	0
1 tsp	Parkay Margarine-70% Veg Oil-Stick	--	1.67	36.67	--	0	0
3.2 oz-wt	Bacon-Svg JBX	--	518.40	2462.40	--	0	0
1 tbs	Mayonnaise-(Soybean Oil) w/Salt	3.86	4.69	78.38	0.02	0	0
0.25 cup	Romaine (Cos) Lettuce-Chopped-Cup	6.30	40.60	1.12	0.04	0	0
2 each	Tomatoes-LrgSlice-Sandwich Component SUB	--	--	2.00	--	0	0
2 piece	Rye Bread-Toasted-Slice-Reg	66.24	87.84	348.00	0.60	0	0
1 cup	Romaine (Cos) Lettuce-Chopped-Cup	25.20	162.40	4.48	0.14	0	0
1 oz-wt	Cheddar Cheese-Shredded	145.15	27.78	176.05	0.88	0	0
1 tbs	Carrots-Raw-Grated-Cup	3.03	22.21	2.41	0.01	0	0
0.5 oz-wt	Cucumber-Raw, Medium FDA	--	24.34	0	--	0	0
2 tbs	Sunflower Seeds FRL	--	--	16.67	--	0	0
0.25 each	California Avocado-Each	18.16	274.20	5.19	0.18	0	0
0.5 each	Large Poached Egg-Each	44.25	30.00	70.00	0.28	0	0
0.8 oz-wt	Bacon-Svg JBX	--	129.60	615.60	--	0	0
3 tbs	Marie's Buttermilk RanchDressing/Dip DFV	--	--	345.00	--	0	0
3 cup	Brewed Tea	7.10	262.85	21.31	0.14	142.08	0
0.5 tsp	Fresh Lemon Juice	0.15	3.15	0.03	0.00	0	0
16 each	Wheat Thins Crackers-Low Sodium	60.48	69.44	80.00	0.33	0	0
2 tbs	Creamy Peanut Butter	118.08	214.08	149.44	0.93	0	0
3 oz-wt	Beef Roast-BottomRound-All-1/4"Trim-Rstd	208.37	239.84	42.53	4.18	0	0
1 cup	Long Grain Brown Rice-Ckd	161.85	83.85	9.75	1.23	0	0
0.25 cup	Broccoli Pieces-Stir Fried	25.70	126.36	10.53	0.16	0	0
0.25 cup	Carrot Slices-Stir Fried	17.16	125.97	13.65	0.08	0	0
2 oz-wt	Med Sweet Green Bell Peppers-Raw-Each	10.77	100.36	1.13	0.07	0	0
0.25 cup	Mushroom Pieces/Slices-Raw-Cup	18.20	64.75	0.70	0.13	0	0
1 cup	2% Fat Milk-Vitamin A Added	231.80	375.76	122.00	0.95	0	0
1 cup	Vanilla Ice Cream	138.60	262.68	105.60	0.91	0	0
	Totals	1691.99	4441.23	5302.36	17.31	279.54	0

Student Name: Exercise#2 - Dietary Fats & Cholesterol
Student ID #:
Instructor Name:
Class Days:
Class Time:

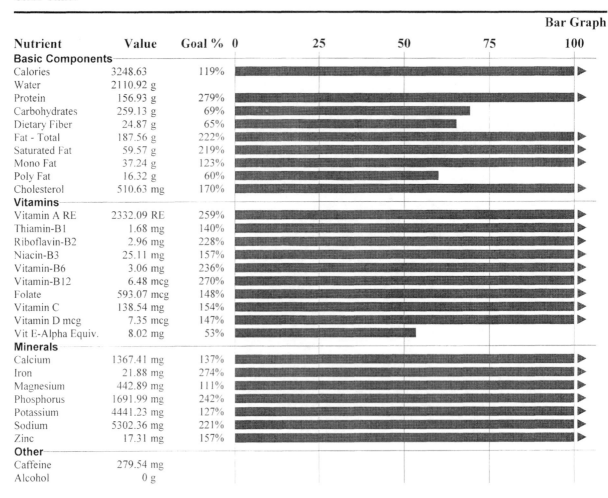

Nutrient	Value	Goal %	Bar Graph
Basic Components			
Calories	3248.63	119%	
Water	2110.92 g		
Protein	156.93 g	279%	
Carbohydrates	259.13 g	69%	
Dietary Fiber	24.87 g	65%	
Fat - Total	187.56 g	222%	
Saturated Fat	59.57 g	219%	
Mono Fat	37.24 g	123%	
Poly Fat	16.32 g	60%	
Cholesterol	510.63 mg	170%	
Vitamins			
Vitamin A RE	2332.09 RE	259%	
Thiamin-B1	1.68 mg	140%	
Riboflavin-B2	2.96 mg	228%	
Niacin-B3	25.11 mg	157%	
Vitamin-B6	3.06 mg	236%	
Vitamin-B12	6.48 mcg	270%	
Folate	593.07 mcg	148%	
Vitamin C	138.54 mg	154%	
Vitamin D mcg	7.35 mcg	147%	
Vit E-Alpha Equiv.	8.02 mg	53%	
Minerals			
Calcium	1367.41 mg	137%	
Iron	21.88 mg	274%	
Magnesium	442.89 mg	111%	
Phosphorus	1691.99 mg	242%	
Potassium	4441.23 mg	127%	
Sodium	5302.36 mg	221%	
Zinc	17.31 mg	157%	
Other			
Caffeine	279.54 mg		
Alcohol	0 g		

1.   Using the Exchange System, estimate the amount of protein in the following meal. First, identify the exchange list and the number of "exchanges" per item. Then, determine the grams of protein per item. After you estimate the protein in this meal, compare your estimate to the Diet Analysis Plus spreadsheet.

1 large flour tortilla (weight 2 oz.)   1 cup shredded lettuce
¾ cup cooked kidney beans (or other beans)   1 cup cooked rice
2 oz. cooked ground beef (regular; 20% fat by weight)   2 tablespoons tomato paste
1 ounce grated Monterey Jack cheese   2 tbsp. sour cream (regular, not nonfat)
½ sliced tomato   1 cup orange juice with 1 cup club soda

Total g of protein you estimated: Total g of protein from spreadsheet:

2.   Refer to the meal above and answer the following questions.

        a.   Identify the <u>main</u> sources of complete protein in the meal.

        b.   What plant foods in this meal are <u>significant</u> sources (2 grams or more per serving) of protein?

        c.   Lysine is a limiting essential amino acid. What plant foods in this meal are low in lysine?

        d.   Identify the plant proteins in this meal that <u>provide</u> lysine.

        e.   Methionine and tryptophan are limiting essential amino acids. What plant foods in this meal are low in these two amino acids?

        f.   Identify the plant proteins in this meal that <u>provide</u> methionine and tryptophan.

        g.   Identify the complementary plant proteins in this menu.

        h.   Are any of the animal proteins in this meal low in lysine, methionine or tryptophan?

3.   Calculate your recommended protein intake (RDA).

Author: Elaine M. Long

4.	What percentage of your RDA protein standard did this meal provide? Show your calculation and label your units.

5.	What is the ratio of grams of animal protein to grams of total protein in this meal? Comment on the ratio (is it high, low or about right).

6.	Summarize your answers to questions 1-5 (total grams of protein, complete/incomplete proteins, your RDA and %RDA for protein, ratio of animal to total protein) about this specific meal.

7.	What are your conclusions about this meal?

8.	What changes would you make to this meal and why?

9	Calculate the percentage of kcal from protein in this meal using the grams of protein supplied and the kcal in the meal (see attached Diet Analysis Plus printout for kcal). Compare this to the average % kcal from protein in US diets.

10.	If you consume 2400 calories and 15% of your kcal come from protein, how many grams of protein is this?

11.	Likewise, if you consume 800 calories and 15% of your kcal come from protein, how many grams of protein is this?

12.	Use your answers to these two calculations to explain why the <u>best</u> way to evaluate your intake of protein is in terms of % RDA for protein supplied instead of % of kcal from protein.

Student Name: Exercise #3 - Protein & Amino Acids
Student ID #:
Instructor Name:
Class Days:
Class Time:

Spreadsheet

Amount	Food Item	Weight (g)	Cals	H2O (g)	Prot (g)	Carb (g)	Fiber (g)
2 oz-wt	Flour Tortilla-10 inch-RTB	56.70	184.28	15.20	4.93	31.53	1.87
0.75 cup	Progresso Red Kidney Beans-Cnd PLB	195.00	165.00	150.89	10.50	30.00	12.00
2 oz-wt	Ground Beef-Patty-(18%Fat)-Broiled-Well	56.70	158.76	29.97	15.99	0	0
1 oz-wt	Monterey Jack Cheese-Shredded	28.35	105.75	11.63	6.94	0.19	0
0.5 each	Fresh Roma Tomatoes-Raw,Red-Each	31.00	6.51	29.07	0.26	1.44	0.34
1 cup	Looseleaf Lettuce-Shredded-Cup	56.00	10.08	52.64	0.73	1.96	1.06
1 cup	White Rice-Enriched-Reg-Ckd w/Salt	158.00	205.40	108.14	4.25	44.51	0.63
2 tbs	Tomato Paste-Canned-without salt-Cup	32.75	26.86	24.17	1.20	6.32	1.34
2 tbs	Sour Cream DGI	30.00	60.00	21.80	1.00	2.00	0
1 cup	Fresh Orange Juice	248.00	111.60	218.98	1.74	25.79	0.50
1 cup	Club Soda	236.80	0	236.56	0	0	0
	Totals	1129.30	1034.23	899.04	47.54	143.74	17.75

Amount	Food Item	Fat (g)	Sat (g)	Mono (g)	Poly (g)	Chol (mg)	A-RE (RE)
2 oz-wt	Flour Tortilla-10 inch-RTB	4.03	0.99	2.14	0.60	0	0
0.75 cup	Progresso Red Kidney Beans-Cnd PLB	0.75	0	--	--	0	0
2 oz-wt	Ground Beef-Patty-(18%Fat)-Broiled-Well	10.00	3.93	4.38	0.37	57.27	0
1 oz-wt	Monterey Jack Cheese-Shredded	8.58	5.41	2.48	0.25	25.23	63.50
0.5 each	Fresh Roma Tomatoes-Raw,Red-Each	0.10	0.01	0.02	0.04	0	19.22
1 cup	Looseleaf Lettuce-Shredded-Cup	0.17	0.02	0.01	0.09	0	106.40
1 cup	White Rice-Enriched-Reg-Ckd w/Salt	0.44	0.12	0.14	0.12	0	0
2 tbs	Tomato Paste-Canned-without salt-Cup	0.18	0.03	0.03	0.07	0	79.91
2 tbs	Sour Cream DGI	5.00	3.50	1.50	0	20.00	40.00
1 cup	Fresh Orange Juice	0.50	0.06	0.09	0.10	0	49.60
1 cup	Club Soda	0	0	0	0	0	0
	Totals	29.75	14.07	10.77	1.66	102.50	358.63

Student Name: Exercise #3 - Protein & Amino Acids
Student ID #:
Instructor Name:
Class Days:
Class Time:

Spreadsheet

Amount	Food Item	B1 (mg)	B2 (mg)	B3 (mg)	B6 (mg)	B12 (mcg)	Fola (mcg)
2 oz-wt	Flour Tortilla-10 inch-RTB	0.30	0.17	2.03	0.03	0	69.74
0.75 cup	Progresso Red Kidney Beans-Cnd PLB	--	--	--	--	--	--
2 oz-wt	Ground Beef-Patty-(18%Fat)-Broiled-Well	0.03	0.14	3.38	0.17	1.54	6.24
1 oz-wt	Monterey Jack Cheese-Shredded	0.00	0.11	0.03	0.02	0.24	5.10
0.5 each	Fresh Roma Tomatoes-Raw,Red-Each	0.02	0.01	0.19	0.02	0	4.65
1 cup	Looseleaf Lettuce-Shredded-Cup	0.03	0.04	0.22	0.03	0	28.00
1 cup	White Rice-Enriched-Reg-Ckd w/Salt	0.26	0.02	2.33	0.15	0	91.64
2 tbs	Tomato Paste-Canned-without salt-Cup	0.05	0.06	1.06	0.12	0	7.21
2 tbs	Sour Cream DGI	--	--	--	--	--	--
1 cup	Fresh Orange Juice	0.22	0.07	0.99	0.10	0	74.40
1 cup	Club Soda	0	0	0	0	0	0
	Totals	0.92	0.63	10.23	0.65	1.78	286.98

Amount	Food Item	Vit C (mg)	D-mcg (mcg)	E-aTE (mg)	Calc (mg)	Iron (mg)	Mag (mg)
2 oz-wt	Flour Tortilla-10 inch-RTB	0	0	0.52	70.88	1.87	14.74
0.75 cup	Progresso Red Kidney Beans-Cnd PLB	0	--	--	60.00	2.70	--
2 oz-wt	Ground Beef-Patty-(18%Fat)-Broiled-Well	0	0.17	0.11	6.80	1.39	13.61
1 oz-wt	Monterey Jack Cheese-Shredded	0	0.06	0.10	211.49	0.20	7.65
0.5 each	Fresh Roma Tomatoes-Raw,Red-Each	5.92	0	0.12	1.55	0.14	3.41
1 cup	Looseleaf Lettuce-Shredded-Cup	10.08	0	0.25	38.08	0.78	6.16
1 cup	White Rice-Enriched-Reg-Ckd w/Salt	0	0	0.07	15.80	1.90	18.96
2 tbs	Tomato Paste-Canned-without salt-Cup	13.89	0	1.41	11.46	0.64	16.70
2 tbs	Sour Cream DGI	0	0	--	40.00	0	--
1 cup	Fresh Orange Juice	124.00	0	0.22	27.28	0.50	27.28
1 cup	Club Soda	0	0	0	11.84	0.02	2.37
	Totals	153.89	0.24	2.80	495.18	10.14	110.88

Student Name: Exercise #3 - Protein & Amino Acids
Student ID #:
Instructor Name:
Class Days:
Class Time:

Spreadsheet

Amount	Food Item	Phos (mg)	Potas (mg)	Sod (mg)	Zinc (mg)	Caff (mg)	Alco (g)
2 oz-wt	Flour Tortilla-10 inch-RTB	70.31	74.28	271.03	0.40	0	0
0.75 cup	Progresso Red Kidney Beans-Cnd PLB	--	--	420.00	--	0	0
2 oz-wt	Ground Beef-Patty-(18%Fat)-Broiled-Well	103.19	197.88	50.46	3.52	0	0
1 oz-wt	Monterey Jack Cheese-Shredded	125.87	22.96	151.96	0.85	0	0
0.5 each	Fresh Roma Tomatoes-Raw,Red-Each	7.44	68.82	2.79	0.03	0	0
1 cup	Looseleaf Lettuce-Shredded-Cup	14.00	147.84	5.04	0.16	0	0
1 cup	White Rice-Enriched-Reg-Ckd w/Salt	67.94	55.30	603.56	0.77	0	0
2 tbs	Tomato Paste-Canned-without salt-Cup	25.87	306.87	28.82	0.26	0	0
2 tbs	Sour Cream DGI	--	--	45.00	--	0	0
1 cup	Fresh Orange Juice	42.16	496.00	2.48	0.12	0	0
1 cup	Club Soda	0	4.74	49.73	0.24	0	0
	Totals	456.79	1374.69	1630.86	6.36	0	0

Student Name: Exercise #3 - Protein & Amino Acids
Student ID #:
Instructor Name:
Class Days:
Class Time:

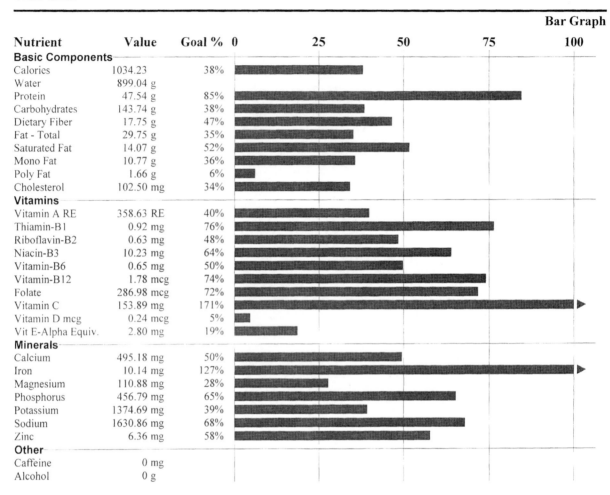

Nutrient	Value	Goal %	Bar Graph
Basic Components			
Calories	1034.23	38%	
Water	899.04 g		
Protein	47.54 g	85%	
Carbohydrates	143.74 g	38%	
Dietary Fiber	17.75 g	47%	
Fat - Total	29.75 g	35%	
Saturated Fat	14.07 g	52%	
Mono Fat	10.77 g	36%	
Poly Fat	1.66 g	6%	
Cholesterol	102.50 mg	34%	
Vitamins			
Vitamin A RE	358.63 RE	40%	
Thiamin-B1	0.92 mg	76%	
Riboflavin-B2	0.63 mg	48%	
Niacin-B3	10.23 mg	64%	
Vitamin-B6	0.65 mg	50%	
Vitamin-B12	1.78 mcg	74%	
Folate	286.98 mcg	72%	
Vitamin C	153.89 mg	171%	
Vitamin D mcg	0.24 mcg	5%	
Vit E-Alpha Equiv.	2.80 mg	19%	
Minerals			
Calcium	495.18 mg	50%	
Iron	10.14 mg	127%	
Magnesium	110.88 mg	28%	
Phosphorus	456.79 mg	65%	
Potassium	1374.69 mg	39%	
Sodium	1630.86 mg	68%	
Zinc	6.36 mg	58%	
Other			
Caffeine	0 mg		
Alcohol	0 g		

EXERCISE #4: EVALUATION OF YOUR PROTEIN INTAKE USING DIET ANALYSIS PLUS

1.   Refer to your **Profile**:

What is your RDA for kcal?   _____ kcal per day

What is your DV for fat?   _____ grams per day

2a.   What is your RDA for protein based on your healthy body weight?

2b.   Now calculate your protein standard based on 12% of your kcal RDA. Compare to 2a.

3.   Using your Diet Analysis Plus Printout (the one with the Bar Graph on it), complete the following table:

	kcal/ %RDA*	grams protein/ % RDA*	grams fat/ %DV*	grams sat fat/ %DV*	milligrams cholesterol/ %DV*
Day 1					
Day 2					
Day 3					
3-Day Average					

Enter both the value (kcal, grams or milligrams) and the % of goal (standard) - draw a diagonal line across the box to separate the two values.

4.   Using the above data:

a.   Looking at your values in the table above, do you find that when your intake of kcal increases, your intakes of protein, fat, saturated fat and cholesterol also increase?

b.   What day did you have the most protein (g/day)? Consider what foods you ate on that day.

c.   Are you surprised at your level of protein intake? Why or why not?

5.   Write the formula for nutrient density here: _____
Calculate the **nutrient density** for protein for Day 1.

6.   Do you have any days where the nutrient density ratio for protein is less than 1? If your answer is yes, what day(s)? **Show your calculation(s).**

7.   For the next calculation, use the day your protein intake was the **highest.** Refer to the **Spreadsheet** for this day and put an "A" by animal sources of protein. You may need to *"guesstimate"* as some items may contain both animal and plant sources of protein (cheese pizza for example). Calculate the ratio of animal protein to total protein in your diet (grams of animal protein divided by total grams of protein). Interpret you answer.

8.   Select one day and identify your **significant sources** of animal protein (>4 grams/serving) and your significant sources of plant protein (>2 grams/serving). Specify the serving size *consumed* and the grams of protein *provided*. This information can be found on your **Spreadsheets**. You can have your computer help you by selecting to print information on a single nutrient. After you enter your food intake for a day, click on **Food list** from the menu bar and pull down the menu, click on **Single nutrient** and then select protein. This information can be printed. Then all you have to do is identify plant or animal. Keep in mind that some foods supply both plant protein and animal protein (a sandwich for example); you will need to estimate amounts accordingly.

Animal Protein Foods	Serving Size Consumed	Grams of Protein	Plant Protein Foods	Serving Size Consumed	Grams of Protein

9.   What percentage of your kcal is supplied by protein, carbohydrate and fat? Compare to recommended percentages. This is listed on your reports under **Source of Calories.**

	% of Kcal from Protein	% of Kcal from Carbohydrate	% of Kcal from Fat
Day 1			
Day 2			
Day 3			
3-Day Average			

1.   Using your Diet Analysis Plus reports, complete the following tables.

Vitamin A   Your RDA for Vitamin A: _____ µg RE

	Retinol Equivalents supplied per day	% Vitamin A RDA supplied	% Kcal RDA supplied	nutrient density ratio *
Day 1				
Day 2				
Day 3				
3-Day Average				

To calculate the nutrient density ratio, divide % nutrient RDA supplied by % kcal RDA supplied. A nutrient density ratio greater than 1 means that your vitamin A intake and your kcal intake are proportional. For vitamin A, it is often possible to supply almost all of it from a single food (which is also often low in calories).

Vitamin C   Your RDA for Vitamin C: _____ mg

	milligrams supplied per day	% Vitamin C RDA supplied	% Kcal RDA supplied	Nutrient density ratio *
Day 1				
Day 2				
Day 3				
3-Day Average				

To calculate the nutrient density ratio, divide % nutrient RDA supplied by % kcal RDA supplied. A nutrient density ratio greater than 1 means that your vitamin C intake and your kcal intake are proportional. For vitamin C, it is often possible to supply almost all of it from a single food.

Author: Elaine M. Long

B-Vitamin   (Select either **thiamin, riboflavin or niacin**):   _____

Your RDA this B-vitamin   _____ mg

	Amount supplied per day (use units)	% Vitamin RDA supplied	% Kcal RDA supplied	nutrient density ratio *
Day 1				
Day 2				
Day 3				
3-Day Average				

To calculate the nutrient density ratio, divide % nutrient RDA supplied by % kcal RDA supplied. A ratio greater than 1 means that your vitamin intake and your kcal intake are proportional. For most B-vitamins, it usually is not possible to meet your nutrient standards when caloric intake is inadequate.

2.   For each vitamin, select one of your three days to analyze. List (by name) your main animal and plant sources for each vitamin; put your sources in order from most to least (you can use Diet Analysis Plus to sort your foods by single nutrients; the printout lists the foods in order of amount supplied from most to least). Specify the serving size **consumed** and the amount of the vitamin **supplied**. Note: some vitamins are mostly supplied by animal foods while others are primarily supplied by plant foods. **Code your food sources as minimally processed/ mostly "natural" (MP) or highly processed (P).**

        Vitamin A   **Day Selected: 1   2   3**   (Circle)

Animal Food	MP or P	Serving Size	Amount Supplied	Plant Food	MP or P	Serving Size	Amount Supplied

Hint - when considering food sources, use your RDA as a reference point. Foods with only 1-2% of your RDA are not significant sources. A significant source supplies 25% of the RDA.

<u>Vitamin C</u>   Day Selected: 1   2   3   (Circle)

Animal Food	MP or P	Serving Size	Amount Supplied	Plant Food	MP or P	Serving Size	Amount Supplied

Hint: Compare amounts to your RDA.

<u>B-Vitamin</u> **(Identify)** _____   Day Selected: 1   2   3   (Circle)

Animal Food	MP or P	Serving Size	Amount Supplied	Plant Food	MP or P	Serving Size	Amount Supplied

Hint: Compare amounts to your RDA or AI.

3.   Using the tables from question 2, compare your food sources of vitamins A, C, and either thiamin, riboflavin or niacin to the significant food sources. For reference, remember that to be considered as a significant source of a nutrient, the food must supply 25% of the RDA in a serving (an excellent source supplies 75% of the RDA and a fair source supplies 10%). Discuss each vitamin in a separate paragraph. **Word process your answer and print on separate pages. Suggested length: 2 pages double-spaced.**

Author: Elaine M. Long

Exercise #6: Vitamin Calculations

Using the following printouts, calculate the following:

1.   Nutrient density ratio for vitamin A.

2.   Nutrient density ratio for vitamin C.

3.   In your own words, state the **meaning of one of the two nutrient density ratios.**

4.   Are there any vitamins with nutrient density ratios less than 1? **If yes, name them. If no, write none.**

5.   What vitamin has the highest nutrient density ratio? The lowest nutrient density ratio?

6.   True or false: this diet is nutrient dense. Explain your answer.

7.   To be considered a significant source of a nutrient a food must supply at least 25% of the RDA/AI; an excellent source must supply at least 75%; a good source supplies 50%; and a fair source supplies 10%.

Author: Elaine M. Long

8. Use the spreadsheet to identify the three highest sources of vitamin E in this person's diet. Her RDA for vitamin E is 15 milligrams alpha-tocopherol. Use this information to classify the three sources (excellent, good, significant or fair).

9. This woman's RDA for folate is 400 µg DFE (Dietary Folate Equivalents). Are there any significant sources (at least 25% of the RDA) of folate in her diet? If yes, list them. *Note: the percentages on the printout represent the % of the total for the specific nutrient supplied by a food not the % of the RDA. Theses percentages are ranked from highest to lowest.*

10. This woman's RDA for vitamin C is 75 milligrams. Are there any excellent sources (at least 75% of the RDA) of vitamin C in her diet? If yes, list them.

11. If you take a supplement with 1 gram of vitamin C, is this considered a mega dose? Calculate how many times the RDA is consumed and compare to the definition of a mega dose. **Approximately how many oranges would this be?**

Student Name: Exercise #6 / #7
Student ID #:
Instructor Name:
Class Days:
Class Time:
Gender: Female
Activity Level: Moderately Active
Height: 5 ft 8 in
Weight: 110 lbs
Age: 27 yrs
BMI: 16.73

Recommended Daily Nutrients

Basic Components

Calories	2003.02	Cholesterol	300.00 mg	Vitamin D mcg	5.00 mcg
Water	-- g	**Vitamins**		Vit E-Alpha Equiv.	15.00 mg
Protein	39.92 g	Vitamin A RE	700.00 RE	**Minerals**	
Carbohydrates	275.42 g	Thiamin-B1	1.10 mg	Calcium	1000.00 mg
Dietary Fiber	28.04 g	Riboflavin-B2	1.10 mg	Iron	18.00 mg
Fat - Total	62.32 g	Niacin-B3	14.00 mg	Magnesium	310.00 mg
Saturated Fat	20.03 g	Vitamin-B6	1.30 mg	Phosphorus	700.00 mg
Mono Fat	22.26 g	Vitamin-B12	2.40 mcg	Potassium	3500.00 mg
Poly Fat	20.03 g	Folate	400.00 mcg	Sodium	2400.00 mg
		Vitamin C	75.00 mg	Zinc	8.00 mg

Student Name: Exercise #6 / #7
Student ID #:
Instructor Name:
Class Days:
Class Time:

Spreadsheet

Amount	Food Item	Weight (g)	Cals	H2O (g)	Prot (g)	Carb (g)	Fiber (g)
2 cup	General Mills Honey Nut Cheerios Cereal	60.00	223.80	1.32	5.40	48.00	3.60
1 cup	Soy Dream Soy Milk-Vanilla-Non Dairy IFI	244.00	147.58	210.43	6.89	21.65	0
1 cup	Water	236.56	0	236.32	0	0	0
1 each	Whole Wheat Bagel (4 1/2" diameter)	110.00	290.81	31.14	12.07	62.03	10.28
1 each	Medium Orange-2 5/8 " Diameter-Each	131.00	61.57	113.64	1.23	15.39	3.14
6 each	Baby Carrots-Raw-Med	60.00	22.80	53.89	0.50	4.90	1.08
4 oz-wt	Veggie Cheese Substitute-Shred-Veget GNF	113.40	240.00	66.08	24.00	8.00	--
1 cup	Crookneck Squash-Bld-Slices-Cup	180.00	36.00	168.66	1.64	7.76	2.52
1 cup	Diced Tomatoes-Cnd DLM	252.00	50.00	235.49	2.00	12.00	4.00
0.25 cup	Carrot Slices-Steamed	39.00	16.81	34.20	0.41	3.94	1.17
1 cup	Navy Beans-Mature Seeds-Ckd w/o Salt	182.00	258.44	114.99	15.83	47.88	11.65
0.25 tsp	Ground Basil	0.35	0.88	0.02	0.05	0.21	0.14
1 piece	Whole Wheat Bread-Slice-Toasted	25.00	69.25	7.50	2.73	12.93	1.85
0.5 oz-wt	Fresh Garlic Cloves-Each Measure	14.18	21.12	8.30	0.90	4.69	0.30
2 cup	Apple Juice-Canned/Bottled,Unsweetened	496.00	233.12	436.13	0.30	57.93	0.30
1 each	LoFatGranola Bar-CrnchyAlmond&BrnSug KLC	21.00	81.90	1.05	1.68	16.38	0.50
2 cup	Water	473.12	0	472.65	0	0	0
	Totals	2637.60	1754.08	2191.82	75.63	323.68	41.53

Student Name: Exercise #6 / #7
Student ID #:
Instructor Name:
Class Days:
Class Time:

Spreadsheet

Amount	Food Item	Fat (g)	Sat (g)	Mono (g)	Poly (g)	Chol (mg)	A-RE (RE)
2 cup	General Mills Honey Nut Cheerios Cereal	2.40	0.48	0.74	0.92	0	300.60
1 cup	Soy Dream Soy Milk-Vanilla-Non Dairy IFI	3.94	0.49	--	--	0	0
1 cup	Water	0	0	0	0	0	0
1 each	Whole Wheat Bagel (4 1/2" diameter)	1.61	0.27	0.25	0.62	0	0
1 each	Medium Orange-2 5/8 " Diameter-Each	0.16	0.02	0.03	0.03	0	26.20
6 each	Baby Carrots-Raw-Med	0.32	0.06	0.02	0.16	0	901.20
4 oz-wt	Veggie Cheese Substitute-Shred-Veget GNF	12.00	0	--	--	0	--
1 cup	Crookneck Squash-Bld-Slices-Cup	0.56	0.12	0.04	0.24	0	50.40
1 cup	Diced Tomatoes-Cnd DLM	0	0	0	0	0	100.00
0.25 cup	Carrot Slices-Steamed	0.07	0.01	0.00	0.03	0	988.65
1 cup	Navy Beans-Mature Seeds-Ckd w/o Salt	1.04	0.27	0.09	0.45	0	0.36
0.25 tsp	Ground Basil	0.01	0.00	0.00	0.01	0	3.28
1 piece	Whole Wheat Bread-Slice-Toasted	1.20	0.26	0.47	0.28	0	0
0.5 oz-wt	Fresh Garlic Cloves-Each Measure	0.07	0.01	0.00	0.04	0	0
2 cup	Apple Juice-Canned/Bottled,Unsweetened	0.55	0.09	0.02	0.16	0	0.50
1 each	LoFatGranola Bar-CrnchyAlmond&BrnSug KLC	1.55	0.23	0.38	0.94	0	149.73
2 cup	Water	0	0	0	0	0	0
	Totals	25.47	2.31	2.04	3.88	0	2520.92

Student Name: Exercise #6 / #7
Student ID #:
Instructor Name:
Class Days:
Class Time:

Spreadsheet

Amount	Food Item	B1 (mg)	B2 (mg)	B3 (mg)	B6 (mg)	B12 (mcg)	Fola (mcg)
2 cup	General Mills Honey Nut Cheerios Cereal	0.75	0.85	10.02	1.00	3.00	400.20
1 cup	Soy Dream Soy Milk-Vanilla-Non Dairy IFI	0.15	0.07	1.18	0.12	--	59.03
1 cup	Water	0	0	0	0	0	0
1 each	Whole Wheat Bagel (4 1/2" diameter)	0.34	0.28	5.75	0.30	0	65.74
1 each	Medium Orange-2 5/8 " Diameter-Each	0.11	0.05	0.37	0.08	0	39.30
6 each	Baby Carrots-Raw-Med	0.02	0.03	0.53	0.05	0	19.80
4 oz-wt	Veggie Cheese Substitute-Shred-Veget GNF	--	--	--	--	--	--
1 cup	Crookneck Squash-Bld-Slices-Cup	0.09	0.09	0.92	0.17	0	36.00
1 cup	Diced Tomatoes-Cnd DLM	--	--	--	--	--	--
0.25 cup	Carrot Slices-Steamed	0.03	0.02	0.34	0.05	0	5.19
1 cup	Navy Beans-Mature Seeds-Ckd w/o Salt	0.37	0.11	0.97	0.30	0	254.80
0.25 tsp	Ground Basil	0.00	0.00	0.02	0.01	0	0.96
1 piece	Whole Wheat Bread-Slice-Toasted	0.08	0.05	0.97	0.05	0.00	9.75
0.5 oz-wt	Fresh Garlic Cloves-Each Measure	0.03	0.02	0.10	0.18	0	0.43
2 cup	Apple Juice-Canned/Bottled,Unsweetened	0.10	0.08	0.50	0.15	0	0
1 each	LoFatGranola Bar-CrnchyAlmond&BrnSug KLC	0.15	0.17	1.99	0.21	0	0
2 cup	Water	0	0	0	0	0	0
	Totals	2.22	1.83	23.67	2.66	3.00	891.19

Amount	Food Item	Vit C (mg)	D-mcg (mcg)	E-aTE (mg)	Calc (mg)	Iron (mg)	Mag (mg)
2 cup	General Mills Honey Nut Cheerios Cereal	12.00	2.10	0.62	199.80	9.00	64.20
1 cup	Soy Dream Soy Milk-Vanilla-Non Dairy IFI	0	--	--	39.35	1.77	39.35
1 cup	Water	0	0	0	4.73	0.02	2.37
1 each	Whole Wheat Bagel (4 1/2" diameter)	0.01	0	0.99	31.89	3.52	115.56
1 each	Medium Orange-2 5/8 " Diameter-Each	69.69	0	0.31	52.40	0.13	13.10
6 each	Baby Carrots-Raw-Med	5.04	0	0.24	13.80	0.47	7.20
4 oz-wt	Veggie Cheese Substitute-Shred-Veget GNF	--	--	--	1000.00	--	--
1 cup	Crookneck Squash-Bld-Slices-Cup	9.90	0	0.22	48.60	0.65	43.20
1 cup	Diced Tomatoes-Cnd DLM	18.00	--	--	40.00	0.72	--
0.25 cup	Carrot Slices-Steamed	2.73	0	0.16	10.53	0.20	5.85
1 cup	Navy Beans-Mature Seeds-Ckd w/o Salt	1.64	0	0.73	127.40	4.51	107.38
0.25 tsp	Ground Basil	0.21	0	0.01	7.40	0.15	1.48
1 piece	Whole Wheat Bread-Slice-Toasted	0	0.05	0.29	20.25	0.93	24.25
0.5 oz-wt	Fresh Garlic Cloves-Each Measure	4.42	0	0.00	25.66	0.24	3.54
2 cup	Apple Juice-Canned/Bottled,Unsweetened	4.46	0	0.05	34.72	1.84	14.88
1 each	LoFatGranola Bar-CrnchyAlmond&BrnSug KLC	0	0	0	7.35	1.81	18.27
2 cup	Water	0	0	0	9.46	0.05	4.73
	Totals	128.10	2.15	3.62	1673.34	26.00	465.36

Student Name: Exercise #6 / #7
Student ID #:
Instructor Name:
Class Days:
Class Time:

| | | | | | | Spreadsheet |
Amount	Food Item	Phos (mg)	Potas (mg)	Sod (mg)	Zinc (mg)	Caff (mg)	Alco (g)
2 cup	General Mills Honey Nut Cheerios Cereal	199.80	183.00	538.80	7.50	0	0
1 cup	Soy Dream Soy Milk-Vanilla-Non Dairy IFI	98.39	255.81	137.74	0.59	0	0
1 cup	Water	0	0	7.10	0.07	0	0
1 each	Whole Wheat Bagel (4 1/2" diameter)	317.24	379.09	591.99	2.52	0	0
1 each	Medium Orange-2 5/8 " Diameter-Each	18.34	237.11	0	0.09	0	0
6 each	Baby Carrots-Raw-Med	22.80	167.40	21.00	0.09	0	0
4 oz-wt	Veggie Cheese Substitute-Shred-Veget GNF	720.00	40.00	1560.00	--	0	0
1 cup	Crookneck Squash-Bld-Slices-Cup	70.20	345.60	1.80	0.70	0	0
1 cup	Diced Tomatoes-Cnd DLM	--	--	320.00	--	0	0
0.25 cup	Carrot Slices-Steamed	17.16	125.97	13.65	0.08	0	0
1 cup	Navy Beans-Mature Seeds-Ckd w/o Salt	285.74	669.76	1.82	1.93	0	0
0.25 tsp	Ground Basil	1.71	12.02	0.12	0.02	0	0
1 piece	Whole Wheat Bread-Slice-Toasted	64.50	70.75	148.00	0.55	0	0
0.5 oz-wt	Fresh Garlic Cloves-Each Measure	21.69	56.84	2.41	0.16	0	0
2 cup	Apple Juice-Canned/Bottled,Unsweetened	34.72	590.24	14.88	0.15	0	0
1 each	LoFatGranola Bar-CrnchyAlmond&BrnSug KLC	52.08	52.29	61.11	0.46	0	0
2 cup	Water	0	0	14.19	0.14	0	0
	Totals	1924.37	3185.88	3434.61	15.06	0	0

Student Name: Exercise #6 / #7
Student ID #:
Instructor Name:
Class Days:
Class Time:

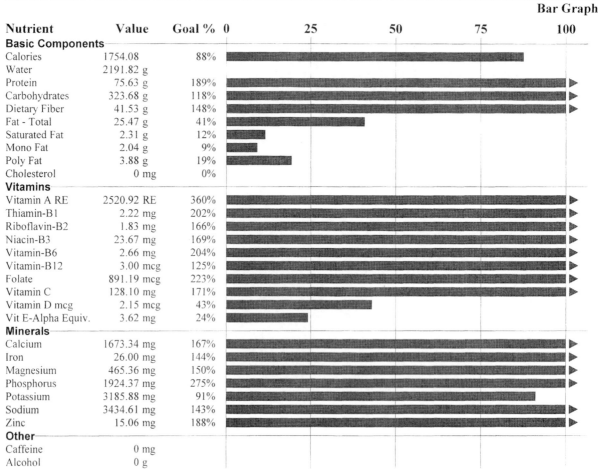

Nutrient	Value	Goal %	Bar Graph
Basic Components			
Calories	1754.08	88%	
Water	2191.82 g		
Protein	75.63 g	189%	
Carbohydrates	323.68 g	118%	
Dietary Fiber	41.53 g	148%	
Fat - Total	25.47 g	41%	
Saturated Fat	2.31 g	12%	
Mono Fat	2.04 g	9%	
Poly Fat	3.88 g	19%	
Cholesterol	0 mg	0%	
Vitamins			
Vitamin A RE	2520.92 RE	360%	
Thiamin-B1	2.22 mg	202%	
Riboflavin-B2	1.83 mg	166%	
Niacin-B3	23.67 mg	169%	
Vitamin-B6	2.66 mg	204%	
Vitamin-B12	3.00 mcg	125%	
Folate	891.19 mcg	223%	
Vitamin C	128.10 mg	171%	
Vitamin D mcg	2.15 mcg	43%	
Vit E-Alpha Equiv.	3.62 mg	24%	
Minerals			
Calcium	1673.34 mg	167%	
Iron	26.00 mg	144%	
Magnesium	465.36 mg	150%	
Phosphorus	1924.37 mg	275%	
Potassium	3185.88 mg	91%	
Sodium	3434.61 mg	143%	
Zinc	15.06 mg	188%	
Other			
Caffeine	0 mg		
Alcohol	0 g		

Bar Graph scale: 0   25   50   75   100

EXERCISE #7: MINERAL CALCULATIONS

Use the Diet Analysis Plus printout that followed Exercise #6 to answer the following questions and complete the calculations. Show your work.

1.	What mineral has the highest nutrient density ratio?

2.	What mineral has the lowest nutrient density ratio?

3.	Calculate the nutrient density ratio for magnesium.

4.	Interpret the nutrient density ratio that you calculated in question 3.

5.	Sodium does not have an RDA but it does have a Daily Value. Show how the %Goal for sodium in this daily intake was calculated.

6.	When evaluating an individual's intake of sodium, a ratio known as milligrams of sodium per kcal is often used. Calculate this ratio using the information on the printout.

7.	Interpret the ratio from question 6. What is the recommended ratio?

8.	Use the information on amount of sodium in foods in this person's intake and make specific suggestions to this person. Use the spreadsheet. **Remember when you use Diet Analysis Plus, you can print information on single nutrients; the printout will rank/ order food sources from highest to lowest.**

9.	This intake supplies 26.00 milligrams of iron with 1754.08 kcal. Express this value in milligrams per 1000 kcal. The usual intake of iron is 5-6 milligrams per 1000 kcal. Using your calculation, how does this diet compare?

10.	Zinc is found primarily in animal flesh and vegetarians may have lower intakes. Look up some vegan food sources of zinc that you might like to eat. Use this information and make specific suggestions to increase the amount of zinc in this person's diet. Use the percent of the RDA for significant and fair sources as your guideline. Include amounts of zinc in your selected foods in your answer.

Author: Elaine M. Long

11. Some soy products are sources of calcium for vegetarians who don't consume dairy products. This person consumed soy milk with her cereal, and ate some casserole containing soy cheese substitute. Using the Adequate Intake guideline for calcium for young adults 19-30, is either food a significant source? Consider % supplied for the serving size. Show your calculations.

12. Using the RDA standard for magnesium for 19-30 year old females as a "yardstick" are there any significant sources of magnesium in this diet? Name them. Show your calculation.

13. Most foods are considered fair sources of iron (supplying approximately 10% of the RDA). What are the four highest sources of iron in this diet? Evaluate these food sources in terms of your personal RDA for iron. Show your calculations.

This person's evening meal included a casserole made from soy cheese substitute, squash, tomatoes, carrots, navy beans, basil and garlic, plus a slice of whole-wheat toast, a granola bar, apple juice, and water (the last 11 items on the spreadsheet). Using these 11 foods, complete questions 14-17.

14. Determine the total iron in this meal **and** the amount of vitamin C in this dinner of 11 food items.

15. Calculate the amount of heme iron **and** non-heme iron in the meal. **Show your calculations.**

16. Is this a high, medium, or low availability meal? **Explain your answer in terms of the amount of vitamin C and the amount of MFP in the meal.**

17. Complete the calculation of absorbable iron. Show your work and circle your answer. What do you conclude about this dinner?

1.   Use your Diet Analysis Plus reports to complete the following tables. This information can be found on your daily summaries and the 3-day average intake.

Calcium   Your AI for Calcium: _____ mg

	milligrams supplied per day	% Calcium AI supplied	% Kcal RDA supplied	nutrient density ratio *
Day 1				
Day 2				
Day 3				
3-Day Average				

To calculate the nutrient density ratio, divide % nutrient AI or RDA supplied by % kcal RDA supplied. A nutrient density ratio greater than 1 means that your calcium intake and your kcal intake are proportional. For calcium, it is generally possible to supply almost all of your RDA from two or three nutrient dense foods.

Iron   Your RDA for Iron: _____ mg

	milligrams supplied per day	% Iron RDA supplied	% Kcal RDA supplied	nutrient density ratio *
Day 1				
Day 2				
Day 3				
3-Day Average				

To calculate the nutrient density ratio, divide % nutrient RDA supplied by % kcal RDA supplied. A ratio greater than 1 means that your iron intake and your kcal intake are proportional. For iron, it is difficult to meet your RDA without consuming adequate kcal.

Author: Elaine M. Long

<u>Sodium</u>   There are several guidelines or standards for evaluating sodium intakes.

Daily Value: _____ mg/day   Milligram sodium per kcal ratio:_____

	milligrams per day	Kcal per day	milligram sodium per kcal ratio*	% DV**
Day 1				
Day 2				
Day 3				
3-Day Average				

*To calculate the milligram sodium per kcal ratio, divide milligrams sodium per day by kcal per day. Current dietary guidelines suggest a sodium intake of 1 - 2 mg. sodium per kcal.
**To calculate %DV divide milligrams sodium per day by the DV and express as a percentage.

2.   Iron availability is calculated for each meal or snack. **Select one meal from your three-day** intake to use for this calculation. Use **a meal** with at least **four** foods. You will need to calculate the values for heme and non-heme iron. Heme iron is only found in animal flesh (MFP); 40% of the total iron in MFP is heme iron. Non-heme iron is found in all plant foods (nuts, seeds, fruits, vegetables, grains, legumes), milk and eggs plus the remaining 60% iron in MFP. A quick way to calculate non-heme iron is to first determine the amount of heme iron in a meal and then subtract this value from the total iron in the meal.

Food	Serving size	Total iron (mg)	Heme iron (mg)	Non-Heme iron (mg)	Vitamin C (mg)
TOTALS	---				

Using **your values from the preceding table**, calculate absorbable iron. First, multiply your milligrams of heme iron by 23% to estimate milligrams of heme iron absorbed:

_____ mg. heme iron x **.23** = _____ mg. heme iron absorbed

Next determine whether your meal is high, medium or low availability for non-heme iron. **This will determine the percentage of non-heme iron absorbed (either 3%, 5%, or 8%).** Multiply your milligrams of non-heme iron by the percentage absorbed:

_____ mg. non-heme iron x _____ absorbed (either **.08, .05 or .03**) = _____ mg. non-heme iron absorbed

_____ mg. heme iron absorbed + _____ mg. non-heme iron absorbed = _____ **total** mg. iron absorbed.

3. For each mineral (calcium, iron, and sodium) select one of your three days to analyze (you may use different days for each mineral). Then select another mineral and analyze one day's intake.

List (name) your main animal and plant sources for each mineral. Specify the serving size consumed and the amount of the mineral supplied. Also identify whether your foods are minimally processed/mostly "natural" foods (milk, yogurt, fruits, vegetables, cereals without added salt and preservatives, most breads, fresh meats, etc.) or highly processed foods (from a "box" or "can" or with added "ingredients"). Use MP for minimally processed/mostly natural foods and P for processed.

Calcium   **Day Selected: 1   2   3**   (Circle)

Animal Food	MP or P	Serving Size	Amount Supplied	Plant Food	MP or P	Serving Size	Amount Supplied

Iron   **Day Selected: 1   2   3**   (Circle)

Animal Food	MP or P	Serving Size	Amount Supplied	Plant Food	MP or P	Serving Size	Amount Supplied

Author: Elaine M. Long

<u>Sodium</u>   **Day Selected: 1   2   3**   (Circle)

Animal Food	MP or P	Serving Size	Amount Supplied	Plant Food	MP or P	Serving Size	Amount Supplied

<u>Selected mineral:</u> _____ (Identify)   **Day Selected: 1   2   3**   (Circle)

Animal Food	MP or P	Serving Size	Amount Supplied	Plant Food	MP or P	Serving Size	Amount Supplied

4. Write a summary of your mineral evaluation. Focus your discussion on your diet, your RDAs/DRIs, and the foods you eat (or should eat more or less of). Support your statements with information from the tables and questions above as well as your computer printouts. Recommend dietary changes to improve your intake of minerals. **Word process your summary and print it on separate pages. Suggested length: 2-3 pages double-spaced.**

Web site: http://www.wadsworth.com/daplus

The following is the screen that will appear. In order to register, you will need a registration code. This will be provided with the purchase of Diet Analysis Plus Online pin code card.

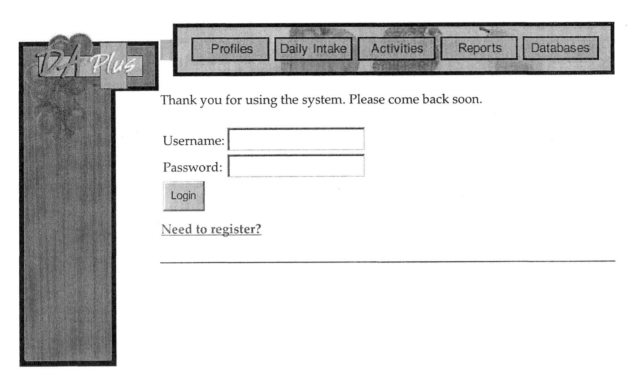

Once you have registered and have a username and password you can login to the web site for DA Plus Online. Write down your username and password (they are case sensitive) and keep this information in a safe and accessible place.

Once you are logged in, you will see a screen with the same headings as above. Click on **Profiles** and enter the information that is requested:
- Birth date
- Gender
- Height
- Weight
- Activity Level (When an item is underlined and you click on it, another screen will come up. In this case, the definitions of each activity level are given.)

Your personal recommendations will be calculated based on your entries. You can edit your profile as you need to. Refer to the left hand side of the screen for the edit command.

The program is relatively simple to use and should be accessible to students of varying degrees of computer literacy. The following instructions are provided as additional information.

Author: Judith S. Matheisz

You are now ready to enter your food intake. Click on **Daily Intake**. You can enter up to 7 days; however, for this assignment, **only enter 3 days**.

You will need to enter **all** foods and beverages (including alcohol and water). Try to include at least **one** weekend day in the three-day record. The days do not have to be consecutive. You can record the food you ate on a Tuesday, Thursday, and Sunday, for example. Record **what** you ate and the **amount**, which includes the **quantity** and **unit of measure**.

When you are ready to start entering your food, type in the name of the food (for example, Harmony cereal) and hit enter. The database will be searched for that food item. Sometimes you will see a message such as: **Sorry, no matches were found for "food, Harmony cereal."** If you just type in **cereal** (more generic) more food items will be found. They are displayed in groups of 10. You will need to advance to the next screen to see more food items. Select a cereal which is closest to the one you ate.

Example of how to enter a food item: Note—It is important to spell correctly; otherwise you will see the same message as above.

Searched for "mayonnaise".
Showing 1 - 10 of 20

Mayonnaise, imit, soybean	+
Mayonnaise, soybean oil, w/salt	+
Mayonnaise, fat free (Kraft General Foods, Inc.)	+
Mayonnaise, low cal, low sod (USDA Survey Database)	+
Salad Dressing/Mayonnaise, Miracle Whip, fat free (Kraft Gen...	+
Mayonnaise, light (Subway International)	+
Mayonnaise (Subway International)	+
Salad Dressing, mayonnaise type, w/salt	+
Salad Dressing/Mayonnaise, Miracle Whip, low cal (USDA Surve...	+
Salad Dressing/Mayonnaise, Miracle Whip, light, cholestfree ...	+

1 2 ☐

If you want to see the nutrition information about the food item, click on the food and another screen will come up. This information may help you decide how to enter the amount of the food you ate. The computer automatically calculates the nutrition information for the amount you will enter.

To add the food item to your food list, select the item and click on the red plus sign.

Then you will need to select which meal you ate the food, the amount (quantity and unit of measure). To enter the meal, use the drop down menu. For amount, first enter the quantity. Quantity must be entered as a whole number, a fraction, or a decimal (0.5, **not** .5). Select the appropriate unit of measure from the drop down menu. The **choose action** is used to edit (update) or delete a food item from your list. Press the enter key or click on the **Enter** sign.

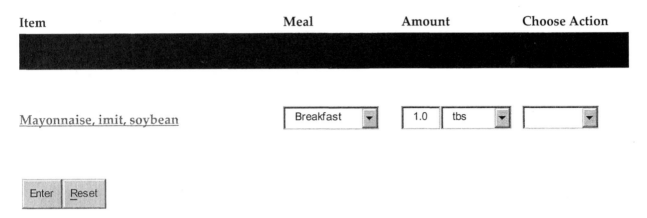

Item	Meal	Amount	Choose Action
Mayonnaise, imit, soybean	Breakfast ▼	1.0	tbs ▼	▼

Enter | Reset

When you have finished entering a day's intake, you can check the nutrition information by clicking on the **Reports** tab. You can click on **All Reports** or you can select which report to view by clicking on the vertical menu on the left side of the screen. You can view the following:
- Bar Graph
- Single Nutrient
- Spreadsheet
- Pyramid
- Ratios/Percents
- Nutrition Facts
- Activity Summary (only if you have entered any activities)

Look at each of the reports to familiarize yourself with data presentation and interpretation.

As you complete each day's intake, be sure to check the entries over very carefully to avoid errors and therefore nonsensical data. Look especially at the total calories eaten. A reasonable amount of calories is in a range of 1500 to 3500 (depending on gender and activity level). If it is not within this range look for possible errors in your food list.

When you have entered all 3 days and have checked for accuracy, it is ready for review by your instructor.

To expedite this review, please send your login information (username and password) for DA Plus Online to your instructor by e-mail.

Author: Judith S. Matheisz

Your instructor will review your food entries online and let you know if any corrections are needed or if you are ready for the Part B, the follow-up analysis.

To print reports, click on the Printer Icon (bottom of left side down menu) to convert to a printer-friendly format. Then input the Print command.

DIET ANALYSIS PLUS ONLINE ASSIGNMENT – PART B
Follow-Up Analysis

This follow-up analysis involves **finding and transferring data** and an **essay**, which will require a thoughtful written analysis of your diet.

- Proceed only after your instructor has reviewed your data and you have made corrections as needed.

- You will need to refer to **all of your daily reports and average reports** of your 3-day diet intake generated by the Diet Analysis Plus Online Version. You will also need to refer to your textbook to answer some of the questions.

- Familiarize yourself with the information on the reports.

  - **Personal Profile** – Gives the recommended amounts based on the personal data you entered into the program – also referred to as the RDA. Please note that for some of the nutrients, the amounts recommended are amounts **NOT** to be exceeded. For example, Fat-Total, types of fat, cholesterol, and sodium.

  - **For each day, Day 1, Day 2, and Day 3**, you will have access to bar graph, single nutrient, spreadsheet, pyramid, ratios/percents, and nutrition facts reports.

  - **Average report** – The program will calculate the average of the days' entries and you will have data similar to each individual day's report.

- You should look at each type of report that is provided so that you can interpret the data. For example, the **bar graph** gives the actual amount you ate (Value) and the Goal % (Value compared to recommended amount expressed as a percent.) Generally you should strive for 100% of Goal **EXCEPT** for fats, cholesterol, and sodium.

- The **single nutrient** data will allow you to view the sources of each component in your daily intake from most to least. For example if you want to see how much of the total fat comes from each food item you entered, you can view that data.

- The **spreadsheet report** breaks down the nutrition information on each food item you entered and adjusts for the amount of food you actually ate.

- The **pyramid** report places the food you ate into the pyramid guide to eating.

- The **ratios/percents** report provides information as to the Source of Calories, Source of Fat, Exchanges, and Ratios.

- The **nutrition facts** report places your entire food intake into the format of a nutrition fact label. It allows you to assess at a glance how you fared relative to the Daily Values.

- You will need your spreadsheet reports for Day 1,2, 3 and your average reports (submit with assignment).

Author: Judith S. Matheisz

Click on REPORTS (menu is across); then click on ALL DAILY REPORTS.

All Daily Reports
Day: 1 2 3 4 5 6 7 Avg

Click on **Avg** first.
Click on the Printer Icon (bottom of left side down menu).
This converts the report to a printer friendly format.
Then input the print command to print the report.

Then click on **Spreadsheets** (left down menu) and click on Day 1.
Click on the Printer Icon (bottom of left side down menu).
Then give the print command to print the report.

Repeat this for days 2 and 3. Select the day, by clicking on the day you want. The day you are on will be in black.

Essay portion of analysis: This will not be accepted in a handwritten format. It must be computer generated (or typewritten). The essay should be at least 2 pages in length (standard margins, double spacing). Points will be deducted for insufficient length. You are expected to demonstrate understanding of nutrition principles as they apply to your personal profile, lifestyle, and dietary practices. You must include responses to the following in your essay.

- Do you feel the recommended calories are high, low or just right for you?
- As you answer this, include information about your age, activity, occupation, health, and fitness level and any other factors that could influence your calorie needs.
- Reflect on the way your lifestyle and perhaps ethnic background or your early family food habits influence your food choices.
- How would you rate the overall quality of your diet? Include data from your 3-day intake to support your rating. It is advisable to first complete the data transfer portion of this assignment before you complete the essay portion.
- What current lifestyle factors contribute to your present diet? Examples could include lack of time, dislike or like of certain foods, interest in healthy eating, etc.
- What dietary changes would you like to achieve?
- What is your position on dietary supplements?
- If you take any dietary supplements such as a multi-vitamin and mineral supplement, what type (brand)? Are the amounts greater than the RDA?
- What are your reasons for taking the supplements?
- Based on the results of your 3-day intake, do you think supplements might be useful for you?
- How has this course and this analysis influenced your dietary practices?

DIET ANALYSIS PLUS ONLINE ASSIGNMENT PART B WORKSHEET

- Attach a copy of the daily average reports and daily spreadsheet reports. (Refer to preceding instructions before proceeding).
- Please provide your username and password for the DA Plus Online program in case the instructor needs to check your reports. Remember the username and password are case sensitive!

Username _____   Password _____

- Attach the essay portion of this assignment.

Refer to the BAR GRAPH of the Average Report and transfer your data below:

	<u>Value</u>	<u>Goal %</u>
Calories		
Protein		
Carbohydrates		
Fat-Total		

Does your calorie intake represent an accurate intake for you? _____

Refer to RATIOS of the Average Report and transfer your data below:

Source of Calories	**%**
Protein	
Carbohydrates	
Fat-Total	
Alcohol	
Total of above	

(total should equal 100% ± 1%)

- New guidelines offer greater flexibility in the mix of fats, carbohydrates, and proteins. The new report suggests that adults can meet their energy and nutrient needs and reduce the risk of developing chronic diseases by eating 20%-35% of calories from fats, 45%-65% from carbohydrates, and 10%-35% from protein. Comment on your % calorie distribution according to the new guidelines.

- Refer to each day's Spreadsheet Report and look at the column heading Carb (g). Sort your carbohydrate food sources as to the **type** of carbohydrate (complex, simple nutritious, concentrated). Only list foods that actually provide at least 15 grams of carbohydrate. List 3 per day.

Complex carbohydrate (mostly starch & fiber rich foods)		
<u>Day 1</u>	<u>Day 2</u>	<u>Day 3</u>

Author: Judith S. Matheisz

Simple nutritious carbohydrate (fruits, milk, and milk products)		
Day 1	**Day 2**	**Day 3**

Simple non-nutritious carbohydrate (sugary foods such as candy, soda pop, desserts)		
Day 1	**Day 2**	**Day 3**

- Which **type** of carbohydrate is present in the greatest amount?

- How does it compare to the recommendation that most of the carbohydrate in the diet should come from complex carbs, followed by simple nutritious, and sparingly from concentrated sugars?

Refer to the BAR GRAPH of the Average Report and transfer your data below:

	Grams	**Goal %**
Dietary Fiber		

- If you did not meet 100% of goal, list at least five (5) specific foods which are fiber rich that you would include in your diet to improve your fiber intake.

Refer to the BAR GRAPH of the Average Report and transfer your data below:

	Grams	**Goal %**
Fat-Total		
Saturated Fat		
Mono Fat		
Poly Fat		
Cholesterol		

Refer to the Ratios of the Average report for Source of Fat and transfer your data:

	%
Saturated Fat	
Mono Fat	
Poly Fat	
Other/Missing	
Total of above	

(should be the same as Fat-Total on the Source of Calories report)

- What type of fat is likely to be in the Other/Missing category?

- Refer to each day's Spreadsheet Report and look at the column heading Prot (g). Categorize your protein food sources into those that are **mostly** from animal foods and those that are **mostly** from plant foods. List 2 per day from each category.

Plant food/protein grams		
Day 1	**Day 2**	**Day 3**

Animal food/protein grams		
Day 1	**Day 2**	**Day 3**

- Which category (plant or animal) provides most of the protein grams in your diet?

- Are the protein food sources in your diet also providing a lot of fat? If so, give some specific examples.

- What changes would you like to make in regard to your protein intake?

Author: Judith S. Matheisz

Refer to the BAR GRAPH of the Average Report and transfer your data below:

- List the vitamins and the Goal % which are equal to or greater than 50% of Goal.

Vitamin	Goal %	Vitamin	Goal %

- List the vitamins and the Goal % which are less than 50% of Goal.

Vitamin	Goal %	Vitamin	Goal %

- For **each vitamin which is less than 50%** of the Goal, research specific foods (refer to your textbook) rich in the vitamin and list foods you would include in your diet to improve your intake.

- List the minerals and the Goal % which are equal or greater than 50% of Goal.

Mineral	Goal %	Mineral	Goal %

- **List the minerals and the Goal % which** are less than 50% of Goal.

Mineral	Goal %	Mineral	Goal %

- For **each mineral which is less than 50%** of the Goal, research specific foods (refer to textbook) rich in the mineral and list foods you would include in your diet to improve your intake.

Refer to each day's Spreadsheet Report and look at the column headings Alco and Caff and transfer your data.

	Alcohol grams	Caffeine Grams
Day 1		
Day 2		
Day 3		

- What are the sources of alcohol in your diet? Is this representative of your usual intake?

- What are the sources of caffeine in your diet? Is this representative of your usual intake?

- What changes would you like to make in regard to alcohol and caffeine intake?

Refer to Pyramid of the Average Report and transfer your data below:

	Recommended servings	Servings consumed
Fats, Oils & Sweets		
Milk, Yogurt & Cheese		
Poultry, Fish, Dry Beans, Eggs & Nuts		
Fruits		
Vegetables		
Bread, Cereal, Rice & Pasta		

Author: Judith S. Matheisz

- What changes would you like to make regarding your distribution of servings in the pyramid?

- What foods contributed to your fats, oils & sweets servings? When interpreting your results for this category, refer back to your actual data. If you did not exceed your fat limit and you do not eat a lot of sugary foods, the number of servings may not be as excessive as you think.

MODULE D: DIET SELF-STUDY EXERCISES

How well do you eat? Our purpose in providing self-study exercises is to encourage you to study your diet. Your reaction to these exercises may be mixed. They will require time and attention, and like your checkbook they have to be done carefully so that they will be accurate and meaningful. The rewards, however, usually outweigh the drawbacks. Most students report that, unlike their checkbooks, these exercises are intriguing, informative, and often reassuring. When doing these self-studies, keep in mind that the more accurate and complete your input, the more reliable the output will be.

Self-Study 1: Record What You Eat and Calculate Your Nutrient Intakes

In this first exercise, you are to record your typical food intake. To get a true average, record all of the meals, snacks, and beverages (including water) you consume for at least two weekdays and one weekend day. If the types and amounts of foods you eat vary greatly from day to day, you may want to record your intake for additional days; make as many copies of the form (Form 1-A if you are hand calculating; Form 1-B if you are using computer software) as you need.

Keep a copy of the form with you throughout the day to record information promptly. As you record each food, make a note of the amount. Weight, measure, or count food carefully. If you are unable to estimate serving sizes, measure out servings the size of a cup, tablespoon, and teaspoon onto a plate or into a bowl to see how they look. In guessing at the sizes of meat portions, it helps to know that a piece of meat the size of the palm of your hand weighs about 3 or 4 ounces. It also helps to know that a slice of cheese (such as sliced American cheese) or a 1 ½-inch cube of cheese weighs roughly 1 ounce. Food labels can also help you estimate serving sizes.

You will also need to carefully describe the foods and how they were prepared. For example, were the peaches you ate fresh or canned? Was the piece of chicken a drumstick or a thigh? Was it roasted or fried?

Nutrition information from many mixed dishes is available, but in some cases you may have to break down mixed dishes to their ingredients. A ham and cheese sandwich, for example, might be listed as 2 slices of bread, 1 tablespoon of mayonnaise, 2 ounces of ham, 1 ounce of cheese, and so on. If you can't discover all the ingredients, estimate the amounts of only the major ones, like the beef, tomatoes, and potatoes in a beef-vegetable soup. You will, of course, make errors in estimating amounts. In calculations of this kind, errors of up to 20 percent are expected and tolerated.

Do not record any nutrient supplements you take. It will be interesting to discover whether your food choices alone deliver the nutrients you need. If they do not, these self-study exercises will enable you to choose foods that do meet your nutrient needs, so that supplements will not be necessary.

Note: The forms for recording food intake and instructions for calculating nutrients differ depending on which method you use to analyze your diet. Forms are found at the end of these Self-Study Exercises. If you are doing these exercises by hand using a food composition table, follow Self-Study 1A (use Forms 1-A, 2-A, and 3-A); if you are using a computer diet analysis program, follow Self-Study 1B (use Form 1-B). In both cases, return to Self-Study 2 to analyze your findings.

Author: Sharon Rady Rolfes

INSTRUCTIONS FOR RECORDING FOOD INTAKE AND HAND CALCULATING NUTRIENT CONTENT USING FOOD COMPOSITION TABLES

Recording. Use one copy of Form 1-A for each day's food intake; make as many copies as needed of the form first. Fill in only columns 1 (food) and 2 (approximate measure or weight) for now. When you have completed your record for all three (or more) days, you can begin looking up each food and beverage in a food composition table and entering the amounts of nutrients for each item in the remaining columns.

If the foods you have eaten are not listed in the food composition table, use the most similar food you can find. For example, if you ate smoked cod (which is not listed), you would not be far off using the values for smoked halibut.

Be careful in recording the nutrient amounts in odd-sized portions. For example, if you used a quarter cup of milk, then you will have to record a fourth of the amount of every nutrient listed for 1 cup of milk. Note the units in which the nutrients are measured:
- Energy is measured in kcalories (kcal).
- Protein, carbohydrate, fiber, fat, and fatty acids are measured in grams (g).
- Cholesterol, calcium, iron, magnesium, phosphorus, potassium, sodium, zinc, thiamin, riboflavin, niacin, vitamin B_6, and vitamin C are measured in milligrams (mg)—thousandths of a gram. Thus 800 milligrams is the same as 0.8 grams calcium; be sure to convert all calcium amounts to milligrams before calculating.
- Folate is measured in micrograms (mcg or μg)—thousandths of a milligram. Thus 400 micrograms folate is the same as 0.4 milligrams folate; be sure to convert all folate amounts to micrograms before calculating.
- Vitamin A is sometimes measured in international units (IU) and sometimes in retinol equivalents (RE): 1 RE equals about 3 IU of vitamin A from animal foods, 10 IU of vitamin A from plant foods, or, on the average, 5 IU (for mixed dishes). Food composition tables may list vitamin A in RE to ease comparison with the RDA, which is also in RE. If you find vitamin A listed in IU on a label, be sure to convert to RE before calculating.

Calculating. Now total the amount of each nutrient you've consumed for each day, and transfer your totals from your three (or more) copies of Form 1-A to Form 2-A. Form 2-A provides a convenient means of deriving and keeping on record an average intake for each nutrient.

Form 2-A also allows you to compare your average intakes with a standard such as the RDA (or AI). Enter the nutrient intakes recommended for a person of your age and sex, referring to your textbook. For intakes of fat, saturated fat, carbohydrate, cholesterol, fiber, and sodium, use the Daily Values. For energy, estimate your daily energy output using the method described in your text. Suspend judgment on the adequacy of your diet for now; the remaining self-studies will guide you in analyzing each of the nutrients provided by your diet.

Use Form 3-A to calculate the percentage of kcalories derived from carbohydrate, fat, protein, and alcohol. (Refer to your textbook if you need help doing the calculations.) Later self-study exercises help you to compare these percentages with recommendations and guidelines.

SELF-STUDY 1-B:
INSTRUCTIONS FOR RECORDING FOOD INTAKE AND CALCULATING NUTRIENT CONTENT USING WADSWORTH'S DIET ANALYSIS PLUS SOFTWARE

Use one copy of Form 1-B for each day's food intake. When you have completed your record for all three (or more) days you can begin entering the foods and beverages into the computer. Items can be entered either by typing in the food name or its code. If using names, be sure to spell correctly so the computer search can find your entry. If using codes, you will first need to search for the foods you have eaten in a food composition table. The far left column of the table provides a computer code number for each food; record this number on your form and use it when entering foods into the computer program. The software documentation included with your computer program will help you to complete your diet analysis using the software.

Author: Sharon Rady Rolfes

FORM 1A

Nutrient Intakes (Use one form for each day.)

Food	Approx Measure or Weight	Energy (kcal)[a]	Prot (g)[a]	Carb (g)[a]	Fiber (g)[a]	Fat (g)[a]	Sat (g)	Mono (g)	Poly (g)	Chol (mg)[a]	Calc (mg)[a]	Iron (mg)[b]	Magn (mg)[a]	Potas (mg)[a]	Sodi (mg)[a]	Zinc (mg)[b]	VitA (RE)[a]	Thia (mg)[b]	VitE (TE)[b]	Ribo (mg)[b]	Niac (mg)[b]	B₆ (mg)[b]	Fol (mu)[a]	VitC (mg)[a]
Totals																								

[a] Compute these values to the nearest whole number.
[b] Compute these values to two decimal places.

FORM 2A

Average Intakes and Comparison with a Standard Intake

Day	Energy (kcal)[a]	Prot (g)[a]	Carb (g)[a]	Fiber (g)[a]	Fat (g)[a]	Sat (g)	Mono (g)	Poly (g)	Chol (mg)[a]	Calc (mg)[a]	Iron (mg)[b]	Magn (mg)[a]	Potas (mg)[a]	Sodi (mg)[a]	Zinc (mg)[b]	VitA (RE)[a]	Thia (mg)[b]	VitE (TE)[b]	Ribo (mg)[b]	Niac (mg)[b]	B_6 (mg)[b]	Fol (mu)[a]	VitC (mg)[a]
1																							
2																							
3																							
Average Daily Intake (divide sum by 3)																							
Recommended Intake[a]																							
Intake as percentage of standard[b]																							

[a] Taken from RDA (or AI) tables, or in the case of fat, carbohydrate, fiber, cholesterol, potassium, and sodium, the Daily Values. For Canadians, use the RDA (AI) tables and the Recommended Nutrient Intakes.

[b] For example, if your intake of protein was 50 grams and the standard for a person your age and sex was 46 grams, then you consumed $(50 \div 46) \times 100$, or 109% of the standard.

FORM 3A
Percentage of kCalories from Protein, Fat, Carbohydrate, and Alcohol

---

Average Daily Intake from Form 2A:

Protein: _____ g/day x 4 kcal/g = (P) _____ kcal/day.

Fat: _____ g/day x 9 kcal/g = (F) _____ kcal/day.

Carbohydrate: _____ g/day x 4 kcal/g = (C) _____ kcal/day.

If you consumed an alcoholic beverage, include its calories.[a]

Alcohol: = (A) _____ kcal/day.

Total kcal/day = (T) _____ kcal/day.

Percentage of kcalories from protein: $(P \div T) \times 100$ = _____% of total kcalories.

Percentage of kcalories from fat: $(F \div T) \times 100$ = _____% of total kcalories.

Percentage of kcalories from carbohydrate: $(C \div T) \times 100$ = _____% of total kcalories.

Percentage of kcalories from alcohol, if any: $(A \div T) \times 100$ = _____% of total kcalories.

Note: The four percentages can total 99, 100, or 101, depending on the way figures were rounded off earlier.

[a] To find out how many kcalories in a beverage are from alcohol, look up the beverage in a food composition table. Figure out how many kcalories are from carbohydrate (multiply carbohydrate grams times 4), fat (fat grams times 9), and protein (protein grams times 4). The remaining kcalories are from alcohol.

FORM 1B
Food Record Input Form

Title of analysis: _____ Date: _____

Name: _____	1. Sedentary: Very inactive, sometimes under someone else's care.
Age: _____	2. Lightly Active: Most office workers and professionals. Equals 8 hours of sleeping, 16 hours of sitting/standing of which 3 hours is light and 1 hour is moderate activity.
Gender:☐ Male ☐ Female – Pregnant	
☐ Female ☐ Female – Lactating	3. Moderately Active: Most persons in light industry, building trades, many farm workers, child care providers, active students, mechanics and commercial fishermen.
Weight: _____ Height: _____	
Activity Level # _____	
(Enter number from choices on right.)	4. Very Active: Full-time athletes, mine or steel workers, unskilled laborers, some agricultural workers, army recruits and soldiers in active service.
	5. Exceptionally Active: Lumberjacks, female construction workers, heavy manual digging.

Code	Amount	Description of Food/Ingredient	Code	Amount	Description of Food/Ingredient

Form 101—92: ESHA Research, Nutrition & Fitness Software, PO Box 13028, Salem, OR 97309
Phone 503-585-6242, Fax 503-585-5543

FORM 1B
Food Record Input Form (continued)

Title of analysis:_____ Date: _____

Code	Amount	Description of Food/Ingredient	Code	Amount	Description of Food/Ingredient

Form 101—92: ESHA Research, Nutrition & Fitness Software, PO Box 13028, Salem, OR 97309
Phone 503-585-6242, Fax 503-585-5543

The remaining self-study exercises can be completed using:

- Forms 2-A and 3-A if you used the hand-calculation method. If you need help with calculations, refer to your textbook.
- The computer printout reports if you used computer software.

SELF-STUDY 2: EVALUATE YOUR ENERGY INTAKE

1. What is your average daily intake?

2. How does your estimated daily output compare with your average energy intake?

3. Have you been gaining or losing weight recently?

If so, is this consistent with the differences between your intake and estimated output?

Author: Sharon Rady Rolfes

4.   If you drank alcoholic beverages, how many drinks did you consume?

How many kcalories did alcohol contribute to your daily energy intake?

What percentage of your energy intake comes from alcohol?

Recommendations suggest that for those who consume alcohol, it should contribute no more than 5 percent of the total energy intake or two drinks daily, whichever is less. Later self-studies examine what percentage of your energy intake comes from protein, fat, and carbohydrate and whether these fall in line with current recommendations.

SELF-STUDY 3: EVALUATE YOUR CARBOHYDRATE INTAKE

1. How many grams of carbohydrate do you consume in an average day? _____

2. How many kcalories does this represent? (Remember, 1 gram of carbohydrate contributes 4 kcalories.) _____

3. It is estimated that you should have at least 125 grams, and ideally much more, of carbohydrate in a day. How does your intake compare with this minimum? _____

    _____

4. What percentage of your total kcalories is contributed by carbohydrate? _____

5. Is your intake in line with the recommendation that 55 to 60 percent of the kcalories in your diet should come from carbohydrate? _____

6. Another dietary goal is that no more than 10 percent of total kcalories should come from refined and other processed sugars and foods high in such sugars. To assess your intake against this standard, sort the carbohydrate-containing food items you ate into three groups:

•   Nutritious foods containing complex carbohydrates (foods such as breads, legumes, and vegetables).

    _____

    _____

•   Nutritious foods containing simple carbohydrates (foods such as milk products and fruits).

    _____

    _____

•   Foods containing mostly concentrated simple carbohydrates (foods such as sugar, honey, molasses, syrup, jam, jelly, candy, cakes, doughnuts, sweet rolls, cola beverages, and so on).

    _____

    _____

Author: Sharon Rady Rolfes

Estimate and include such sources as the syrup of canned fruit, the sugars of flavored yogurts, and the sugars added during processing.

7.   How many grams of carbohydrate did you consume in each of these three categories? _____

_____

How many kcalories (grams times 4)? _____

What percentage of your total kcalories comes from concentrated sugars? _____

Does your concentrated sugar intake fall within the recommended maximum of 10 percent of

total kcalories? _____ If not, what food choices account for the excess sugar? _____

_____

8.   Estimate how many pounds of sugar (concentrated simple carbohydrate) you eat in a year (1

pound = 454 grams). How does your yearly sugar intake compare with the estimated U.S.

average of about 45 pounds per person per year? _____

_____

9.   How many grams of fiber do you consume in an average day? _____

How does your intake compare with the recommendations to consume 20 to 35 grams of dietary

fiber per day? _____

_____

Self-Study 4: Evaluate Your Fat Intake

1. How many grams of fat do you consume on an average day?

2. How many kcalories does this represent? (Remember, 1 gram of fat contributes 9 kcalories.)

3. What percentage of your total energy is contributed by fat?

4. How does your fat intake compare with the recommendation that says fat should contribute not more than 30 percent of total food energy? _____ If it is higher, look over your food records: what specific foods could you cut down on or eliminate, and what foods could you replace them with, to bring your total fat intake into line?

5. How much linoleic acid do you consume? (Refer to your average for polyunsaturated fatty acids, and assume that most of these fatty acids are linoleic acid.)

    Remembering that linoleic acid is a lipid (energy value, 9 kcalories per gram), calculate the number of kcalories it gives you. What percentage of your total energy comes from linoleic acid? A guideline recommends 1 to 3 percent of total energy intake.

Author: Sharon Rady Rolfes

6. The Committee on Dietary Reference Intakes has not established an RDA for omega-3 fatty acids. However, you can guess at the adequacy of your intake by answering the following questions. Do you eat leafy vegetables, fish, and seafood, or other foods listed as sources of omega-3 fatty acids in your text? _____ Do you use canola or soybean oil for home cooking and for salads? _____ If you include just one of these categories of foods each day, you may receive enough omega-3 fatty acids. If you never eat these foods, you might want to find ways to include them.

7. How much cholesterol do you consume daily? _____ How does your cholesterol intake compare with the suggested limit of 300 milligrams a day?

If your intake is high, what foods could you cut down on or eliminate to bring your cholesterol intake within suggested limits?

SELF-STUDY 5: EVALUATE YOUR PROTEIN INTAKE

1.   How many grams of protein do you consume on an average day?

2.   How many kcalories does this represent? (Remember, 1 gram of protein contributes 4 kcalories.)

3.   What percentage of your total food energy is contributed by protein?

4.   Diets that meet the suggested balance of about 55 to 60 percent of the kcalories from carbohydrate, no more than 30 percent from fat, contribute about 10 to 15 percent of total food energy from protein. How does your protein intake compare with this recommendation?

     If your protein intake is out of line, what foods could you consume more of—or less of—to bring it into line?

5.   Calculate your protein RDA (0.8 grams per kilogram of body weight).

     Is it similar to the RDA for an "average" person of your age and sex as shown in the RDA tables?

Author: Sharon Rady Rolfes

6.   Compare your average daily protein intake with your RDA. The *Diet and Health* report suggests that you eat no more than twice your RDA for protein. Does your intake exceed twice your RDA?

        If so, you are spending protein prices for an energy-yielding nutrient and displacing other foods. What substitutions could you make in your day's food choices so that you would derive the energy you need from carbohydrate rather than from protein?

7.   How many of your protein grams are from animal, and how many from plant, foods?

        Assuming that the animal protein is all of high quality, no more than 20 percent of your total protein need come from this source. Should you alter the ratio of plant to animal protein in your diet?

        If you did, what effect would this have on the total *fat* content of your diet?

8.   How is your protein intake distributed through the day? (At what times do you eat how many grams of protein?)

SELF-STUDY 6: EVALUATE YOUR WEIGHT AND HEALTH RISKS

What weight is appropriate for you? When physical health alone is considered, a wide range of weights is acceptable for a person of a given height. Within the safe range, the choice of a weight is up to the individual.

1. Determine whether your current weight is appropriate for your height.

    - Record your height: _____ in (or cm).

    - Record your weight: _____ lb (or kg).

    Look up the acceptable weight range for a person of your height.

    - Record the entire range: _____ to _____ lb (or kg).

    Does your current weight fall within the suggested range? _____ Calculate your BMI using the equation provided in your textbook.

    - Record your BMI: _____ kg/m².

    Look up the risk of disease and mortality for a person with your BMI value.

    - Record your risk level based on your BMI: _____

    If this level of risk is unacceptable, calculate the weight needed for a desired BMI value (divide the desired BMI by the appropriate height factor). For example, a 165-pound person who is 5 feet 5 inches tall has a BMI of 27.5. To obtain a BMI of 22, the person would need to weigh about 133 pounds (22 ÷ 0.166).

    - Record your desired weight based on your height and desired BMI: _____

If you are underweight and your BMI is below 19, you may need to gain weight for your health's sake. If your weight is over the acceptable weight range and your BMI value is associated with an unacceptable risk of disease, you may want to examine your body's fat distribution.

Author: Sharon Rady Rolfes

2.   Determine whether your fat distribution is associated with health risks.

   •   Record your waist measurement: _____ in (or cm).

   Calculate the waist-to-hip ratio by dividing the number of inches (or centimeters) around your waistline by the number of inches (or centimeters) around your hips.

Women with a circumference of greater than 35 inches and men with a circumference of greater than 42 inches are at a high risk for obesity-related health problems.

3.   Check your health history. A family or personal medical history of diabetes, hypertension, or high blood cholesterol signals the need to pay attention to diet and exercise habits. Based on these three considerations, how does your current weight compare with standards that are compatible with health?

SELF-STUDY 7: EVALUATE YOUR VITAMIN INTAKES

1. Compare your average intake with your recommended intake of thiamin. What percentage of your recommended intake of thiamin did you consume?

    Was this enough?

    What foods contribute the greatest amount of thiamin to your diet?

2. Answer these same questions for riboflavin, niacin, vitamin B₆, folate, vitamin C, vitamin A, and

    vitamin E.

Author: Sharon Rady Rolfes

3.   Regarding niacin, remember that preformed dietary niacin is not the only source your body uses; it also uses the amino acid tryptophan, if extra is available after protein needs are met. Did you consume enough niacin preformed in foods to meet your recommended intake?

      If not, calculate your niacin equivalents. Did you consume enough extra protein to bring your intake up to the recommendation for niacin?

4.   Food composition data is often lacking for vitamins D and K, but you can guess at the adequacy of your intake. For vitamin D, answer the following questions: Do you drink fortified milk (read the label)?

      Eat eggs?

      Liver?

      Are you in the sun enough to promote vitamin D synthesis? (Remember, though, that excessive exposure to sun can cause skin cancer in susceptible individuals.)

5.   For vitamin K, does your diet include 2 cups of milk or the equivalent in milk products every day?

      Does it include leafy vegetables frequently (every other day)?

      Do you take antibiotics regularly (which inhibit the production of vitamin K by your intestinal bacteria)?

SELF-STUDY 8: EVALUATE YOUR WATER AND MINERAL INTAKES

1. How much water do you drink daily? _____ How does this compare with the
    recommendation to drink at least 1 milliliter per kcalorie expended? _____
    _____

2. Compare your average intake of calcium with your recommended intake. What percentage of
    your recommended intake of calcium did you consume? _____ Was this enough? _____
    What foods contribute the greatest amount of calcium to your diet? _____
    _____ If you consumed more than
    the recommendation, was this too much? _____ Why or why not? _____
    _____ In what ways would
    you change your diet to improve it in this respect? _____
    _____

3. Answer these same questions for magnesium and potassium. _____
    _____
    _____
    _____
    _____

4. Compare your average intake of sodium with your Daily Value for sodium. What foods
    contribute the greatest amount of sodium to your diet? _____
    _____ If you
    consumed more than the recommendation, how could you change your diet to improve it in this
    respect? _____

5. Compare your average intake of iron with your recommended intake. What percentage of your
    recommended iron intake did you consume? _____ Which of the foods you eat supply the
    most iron? _____ Rank
    your top five iron contributors. _____
    _____
    How many were meats? _____ Legumes? _____ How much of a contribution does
    enriched or whole-grain bread or cereal make to your iron intake? _____ Does your diet

Author: Sharon Rady Rolfes

include refined bread/cereal products, such as pastries, that you could replace with enriched or whole-grain products to increase your iron intake? _____

6. Compute your iron absorption from a meal of your choosing. _____ The RDA assumes you will absorb 10 percent of the iron you ingest. What percentage did you absorb? _____ If you are a man of any age or a woman over 50, you need to absorb about 1 milligram per day; if you are a woman 11 to 50 years old, 1.5 milligrams. How could you best eat to improve your iron absorption? _____

_____

7. Compare your average intake of zinc with your recommended intake. What percentage of your recommended zinc intake did you consume? _____ Which were your best food sources? _____ What guidelines can help you to obtain enough zinc from the foods you eat? _____

_____

8. Food composition data is often lacking for iodine and fluoride, but you can guess at the adequacy of your intake. Are you in an area of the country where the soil is iodine-poor? _____ If so, do you use iodized salt? _____ Is the water in your area fluoridated? (Call the county health department.) _____ If not, how do you and your family ensure that your intakes of fluoride are optimal? _____

_____

9. Review your three-day food record, and separate the foods you ate into two categories: natural, unprocessed foods; and highly processed foods, such as frozen dinners and breakfast bars. Beside each food, record its kcalorie value. What percentage of your energy intake came from highly processed foods? _____ What do you suppose this estimate implies about your diet's mineral adequacy? _____

_____

THINKING THROUGH MY DIET
Exercise #1: Making Food Choices

We decide what to eat, when to eat, and even whether to eat for a variety of reasons. Examine the factors that influence your food choices by keeping a food diary for 24 hours. Record the times and places of meals and snacks, the types and amounts of foods eaten, and a description of your thoughts and feelings when eating. Now examine your food record and consider your choices.

1. Which, if any, of your food choices were influenced by emotions (happiness, boredom, or disappointment, for example)?
2. Was social pressure a factor in any food decisions?
3. Which, if any, of your food choices were influenced by marketing strategies or food advertisements?
4. How large a role do availability, convenience, and economy play in your food choices?
5. Do your age, ethnicity, or health concerns influence your food choices?
6. How many times did you eat because you were truly hungry? How often did you think of health and nutrition when making food choices? Were those food choices different from others made during the day?

Compare the choices you made in your 24-hour food diary to the food guide pyramid.

Food Groups	Suggested Servings	Servings Consumed
Bread, cereal, rice, and pasta	6 to 11 servings	
Vegetable	3 to 5 servings	
Fruit	2 to 4 servings	
Milk, yogurt, and cheese	2 to 3 servings	
Meat, poultry, fish, dry beans, eggs, and nuts	2 to 3 servings	
Fats, oils, and sweets	Use sparingly	

7. Do you eat at least the minimum number of servings from each of the five food groups daily?
8. Do you try to vary your choices within each food group from day to day?
9. What dietary changes could you make to improve your chances of enjoying good health?

Author: Sharon Rady Rolfes

Digestion transforms the foods we eat into nutrients and absorption moves nutrients from the GI tract into the blood. Optimal digestion and absorption depends on the good health of the digestive tract, which is affected by such lifestyle factors as sleep, physical activity, state of mind, and the meals you eat. Identify which of these foods and food habits promote or impede healthy digestion and absorption.

Foods and Food Habits	Promote	Impede
Take small bites of food.	☐	☐
Chew thoroughly before swallowing.	☐	☐
Exercise immediately after eating to prevent weight gain.	☐	☐
Eat a low-fiber diet.	☐	☐
Drink plenty of fluids.	☐	☐
Eat a few large meals instead of several smaller ones.	☐	☐
Eat quickly and then lie down to rest.	☐	☐
Create a meal using citrus fruits and meat.	☐	☐
Tackle family problems at the dinner table.	☐	☐

1.   Do you experience GI distress regularly?

2.   What changes can you make in your eating habits to promote GI health?

THINKING THROUGH MY DIET
Exercise #3: Carbohydrates

Most of the energy we receive from foods comes from carbohydrates. Healthy choices provide complex carbohydrates or naturally occurring simple carbohydrates, rich in water-soluble vitamins and dietary fiber. A diet that is consistently low in dietary fiber and high in added sugar can lead to health problems. Look at these examples of related foods and identify which are most similar to your food choices.

High in fiber/ low in added sugar	Intermediate	Low in fiber/ high in added sugar
Apple with peel	Applesauce, sweetened	Fruit drink, 10% apple juice
Brown rice	Cream of rice cereal	Rice crispy treat
Pumpernickel bread	Bagel, plain	Danish pastry
Baked sweet potato	Candied sweet potato casserole	Sweet potato pie
Corn on the cob	Creamed corn	Frosted corn flakes
Oatmeal	Granola	Granola breakfast bar

1. Do you select whole-grain products and fresh fruits and vegetables regularly?

2. Do you choose foods that increase your intake of fiber and limit your intake of sugars?

Author: Sharon Rady Rolfes

Fats give foods their flavor, texture, and palatability. Unfortunately, these same characteristics entice people to eat too much from time to time. Do you know how to select low-fat foods that will help you meet dietary fat recommendations? Look at these examples of foods and consider how often you select the item that is lower in fat.

Which of these pairs are you most likely to select--

Peanuts	or	pretzels?
Hot dog	or	turkey sandwich?
Whole milk	or	low-fat milk?
Fried chicken	or	baked chicken?
Tuna packed in oil	or	tuna packed in water?
Spaghetti with alfredo sauce	or	with marinara sauce?
Croissants	or	bagels?
Sausage pizza	or	mushroom pizza?

The second item in each pair is lower in fat and making such fat-free or low-fat food choices regularly can help you meet dietary fat recommendations. In addition, eating plenty of whole-grain products, fresh vegetables, legumes, and fruits daily will help to keep your fat intake under control.

Most people in the United States and Canada receive more protein than they need. This is not surprising considering the abundance of food eaten and the central role meats hold in the North American diet. Keep a food diary for one day and then estimate your protein intake for that day. Multiply the number of servings you consumed by the estimated protein per serving to guesstimate your total protein intake.

Food groups	Servings consumed	Estimated protein	Totals
Bread, cereal, rice, and pasta		3 grams/serving	
Vegetable		2 grams/serving	
Fruit		0 grams/serving	
Milk, yogurt, and cheese		8 grams/serving	
Meat, poultry, fish, dry beans, eggs and nuts		7 grams/ounce	
Fats, oils, and sweets		0 grams/serving	
Total estimated protein intake			

The protein RDA for young adults (19 to 24 years old) is 46 grams for women and 58 grams for men. Health experts advise people to maintain moderate protein intakes—between the RDA and twice the RDA.

1.   Do you receive enough, but not too much, protein daily?

2.   How often do you select plant-based protein foods?

Author: Sharon Rady Rolfes

Metabolism explains how the cells in the body use nutrients to meet its needs. Cells may start with small, simple compounds and use them as building blocks to form larger, more complex structures (anabolism). These anabolic reactions involve doing work and so require energy. Alternatively, cells may break down large compounds into smaller ones (catabolism). Catabolic reactions usually release energy. Determine whether the following reactions are anabolic or catabolic.

	Anabolic	Catabolic
A cracker becomes glucose.	☐	☐
Glucose becomes glycogen.	☐	☐
You consume more energy than your body expends.	☐	☐
Fasting.	☐	☐
A piece of ham becomes amino acids.	☐	☐
Amino acids become your muscles.	☐	☐
A cookie becomes fatty acids.	☐	☐
Fatty acids become body fat.	☐	☐
Fatty acids provide energy.	☐	☐

THINKING THROUGH MY DIET
Exercise #7: Overweight, Underweight, and Weight Control

Does your BMI fall between 18.5 and 24.9? If so, you may want to maintain your weight. If not, you may need to gain or lose weight to improve your fitness and health. Determine whether these food and activity choices are typical of your lifestyle.

Food and activity choices	Frequency per week
Promote weight gain:	
Drink plenty of juice.	
Eat energy-dense foods.	
Eat large portions.	
Eat peanut butter crackers between meals.	
Eat three or more large meals a day.	
Promote weight loss:	
Drink plenty of water.	
Eat nutrient-dense foods.	
Eat slowly.	
Eat small portions.	
Limit snacks to healthful choices.	
Limit television watching.	
Participate in physical activity.	
Select low-fat foods.	
Share a restaurant meal or take home leftovers.	

- On the average, do your lifestyle choices promote weight gain, weight loss, or weight maintenance?

Author: Sharon Rady Rolfes

A diet that offers a variety of foods from each group, prepared with reasonable care, serves up ample vitamins. The cereal and bread group delivers thiamin, riboflavin, niacin, and folate. The fruit and vegetable groups excel in folate, vitamin C, vitamin A, and vitamin K. The meat group serves thiamin, niacin, vitamin B_6, and vitamin B_{12}. The milk group stands out for riboflavin, vitamin B_{12}, vitamin A, and vitamin D. Even the miscellaneous group with its vegetable oils provides vitamin E. Determine whether these food choices are typical of your diet.

Food choices	Frequency per week
Citrus fruits	
Dark green, leafy vegetables	
Deep yellow or orange fruits or vegetables	
Legumes	
Milk and milk products	
Vegetable oils	
Whole or enriched grain products	

1.   Do you eat dark green, leafy or deep yellow vegetables daily?

2.   Do you drink vitamin A- and D-fortified milk regularly?

3.   Do you use vegetable oils when you cook?

4.   Do you choose whole or enriched grains, citrus fruits, and legumes often?

The two minerals most likely to fall short in the diet are iron and calcium. Interestingly, both are found in protein-rich foods, but not in the same foods. Meats, fish, and poultry are rich in iron but poor in calcium. Conversely, milk and milk products are rich in calcium but poor in iron. Including meat or meat alternates for iron and milk and milk products for calcium can help defend against iron deficiency and osteoporosis, respectively. Determine whether these food choices are typical of your diet.

Food choices	Frequency per week
Calcium-fortified foods (such as corn tortillas, tofu, cereals, or juices)	
Dark green vegetables (such as broccoli)	
Iron-fortified foods (such as breads or cereals)	
Legumes (such as pinto beans)	
Meats, fish, poultry, or eggs	
Milk or milk products	
Nuts (such as almonds) or seeds (such as sesame seeds)	
Small fish (such as sardines) or fish canned with bones (such as canned salmon)	
Whole or enriched grain products	

1.   Do you eat a variety of foods, including some meats, seafood, poultry, or legumes, daily?

2.   Do you drink at least 3 glasses of milk—or get the equivalent in calcium—every day?

Author: Sharon Rady Rolfes

Fitness depends on a certain minimum amount of physical activity. Ideally, the quantity and quality of the physical activity you select will improve your cardiorespiratory endurance, body composition, strength, and flexibility. Examine your activity choices by keeping an activity diary for one week. For each physical activity, be sure to record the type of activity, the level of intensity, and the duration. In addition, record the times and places of beverage consumption and the types and amounts of beverages consumed. Now compare the choices you made in your one-week activity diary to the guidelines for physical fitness.

1.   How often were you engaged in aerobic activity to improve cardiorespiratory endurance? Was the intensity of aerobic activity between 55 and 90 percent of your maximum heart rate? Did the duration for each session last at least 20 minutes?

2.   How often did you participate in resistance activities to develop strength? Was the intensity enough to enhance muscle strength and improve body composition? Did you perform 8 to 10 different exercises, repeating each one 8 to 12 times?

3.   How often did you stretch to improve your flexibility? Was the intensity enough to develop and maintain a full range of motion? Did you hold each stretch 10 to 30 seconds and repeat each stretch at least four times?

4.   Do you drink plenty of fluids daily, especially water, before, during, and after physical activity?

5.   What changes could you make to improve your fitness?